Studies in Computational Intelligence

Volume 510

Series Editor

Janusz Kacprzyk, Warsaw, Poland

For further volumes:
http://www.springer.com/series/7092

Marcin Mrugalski

Advanced Neural Network-Based Computational Schemes for Robust Fault Diagnosis

Springer

Marcin Mrugalski
Institute of Control and Computation Engineering
Faculty of Electrical Engineering, Computer Science and Telecommunications
University of Zielona Góra
Zielona Góra
Poland

ISSN 1860-949X ISSN 1860-9503 (electronic)
ISBN 978-3-319-03286-3 ISBN 978-3-319-01547-7 (eBook)
DOI 10.1007/978-3-319-01547-7
Springer Cham Heidelberg New York Dordrecht London

Softcover re-print of the Hardcover 1st edition 2014

Printed on acid-free paper

Springer is part of Springer Science+Business Media (www.springer.com)

To my beloved and wonderful wife Beata

Foreword

The artificial neural networks belonging to dynamically developed soft computing methods are widely applied in several modern scientific fields. These advanced computational methods due to their unique properties are used to explore the fields of pattern recognition, signal processing, financial forecasting, modelling and identification, process monitoring and fault diagnosis, fault tolerant control and biomedical engineering applications. Among sources of such popularity of artificial neural networks the ability of modeling of non-linear systems, learning, generalization and adaptation possibilities should be mentioned. Maturity of artificial neural networks, confirmed by numerous publications during last three decades, causes that they are frequently applied. Today such methods are intensively developed in order to solve problems appearing in the dynamically developing fields of science and industry. This fact is confirmed by several practical applications and results of research presented at scientific conferences. One of the fields of applications of artificial neural networks is reliability and safety of technical processes and systems. During the last decade, an intensive research was made on adaptation of classical control and fault detection approaches to the robust fault diagnosis and the fault tolerant control.

The subject matter of this book is devoted to problems of adaptation of artificial neural networks to robust fault diagnosis schemes. This new book presents a neural networks-based modelling and estimation techniques used for designing robust fault diagnosis schemes for non-linear dynamic systems. A part of the book focuses on fundamental issues such as architectures of dynamic neural networks, methods for designing of neural networks and fault diagnosis schemes as well as the importance of robustness. The book is of a tutorial value and can be perceived as a good starting point for the new-comers to this field. The book is also devoted to advanced schemes of description of neural model uncertainty. In particular, the methods of computation of neural networks uncertainty with robust parameter estimation are presented. Moreover, a novel approach for system identification with the state-space

GMDH neural network is delivered. Furthermore, an interesting actuators robust fault detection and isolation schemes are presented. Such approach constitutes the connection of the neural network-base model and robust unknown input filter. All the concepts described in this book are illustrated by both simple academic illustrative examples and practical applications, e.g., DAMADICS benchmark, tunnel furnace diagnosis.

Because of the fact that both theory and practical applications are discussed, the book is useful for both researchers in academia and professional engineers working in the industry. The first group may be especially interested in the fundamental issues and some inspirations regarding future research directions concerning artificial neural networks and robust fault diagnosis. The second group can find practical implementations which may be very helpful in industrial applications of the techniques described in this book. Thus, the book can be strongly recommended to researchers and practitioners in the wide field of fault detection, supervision and safety for technical processes.

Częstochowa, May 2013

Leszek Rutkowski
President of the Polish Neural Network Society

Preface

The quality of models of systems and processes determines the effectiveness of numerous contemporary technical systems i.e., control systems, fault diagnosis systems or fault tolerant control systems. Over the last few decades the scientists and engineers intensively look for the efficient modeling methods of the systems and processes and this book joins into the mainstream of such researches. The subject matter of this book is devoted to artificial neural networks, which thanks to the unique properties are widely applied in non-linear dynamic system identification tasks. In particular, the problem of obtaining neural model uncertainty is deeply considered. The knowledge about the model uncertainty is necessary to design robust fault diagnosis systems. In this book new methods of neural models uncertainty description with the application of the Bounded-Error Approach class of algorithms, Unscented Kalman Filter and Unknown Input Filters are proposed. Such approaches together with the Group Method of Data Handling neural networks allow to develop a novel robust fault detection and isolation methods, which are based on an adaptive threshold approach.

The book results from my research in the area of artificial neural networks and robust fault diagnosis that has been conducted since 1999. This book is primarily a research monograph which presents, in a unified framework, some recent results on the fault diagnosis of the non-linear dynamic systems. The book is intended for researchers, engineers and advanced postgraduate students in the computer science and electrical engineering. Some of the research results presented in this book were developed with the kind support of the National Science Centre in Poland under the grant NN514678440 Predictive fault-tolerant control for non-linear systems.

I wish to express my deepest gratitude to Prof. Józef Korbicz, Ph.D., D.Sc., corresponding member of the Polish Academy of Sciences for suggesting the problem and providing encouraging and inspiring working conditions. I would like to express my sincere thanks to Assoc. Prof. Marcin Witczak, Ph.D., D.Sc, for inspiring a long lasting and successful cooperation. Moreover, I wish to thank my colleagues at the Institute of Control and Computation Engineering of the University of Zielona Góra, who helped me in many, many ways while I was preparing the material contained in this book.

Finally, I would like to express my sincere gratitude to my wife Beata, my son Kamil and family for their love, support, patience and many other important issues of life you have provided to me.

Zielona Góra, May 2013 Marcin Mrugalski

Contents

List of Figures

List of Tables

Abbreviations

AIC	Akaike Information Criterion
ANNs	Artificial Neural Networks
ARS	Adaptive Random Search
BEA	Bounded-Error Approach
BP	Back-Propagation
EKF	Extended Kalman Filter
FDI	Fault Detection and Isolation
FIR	Finite Impulse Response
FPE	Final Prediction Error
FTC	Fault-Tolerant Control
GMDH	Group Method of Data Handling
IIR	Infinite Impulse Response
LMS	Least-Mean Square
MIMO	Multi-Input and Multi-Output
MISO	Multi-Input Single-Output
MLP	Multi-Layer Perceptron
NLMS	Non-linear Least-Mean Square
OBD	Optimal Brain Damage
OBE	Outer Bounding Ellipsoid
OBS	Optimal Brain Surgeon
PDF	Probability Density Function
RTRN	Real Time Recurrent Network
RUIF	Robust Unknown Input Filter
SSM	Soft Selection Method
UIF	Unknown Input Filter
UKF	Unscented Kalman Filter
ZBA	Zonotope-Based Algorithm

1

Introduction

1.1 Introductory Background

The complexity and reliability demands of contemporary industrial systems and technological processes require the development of new Fault Detection and Isolation (FDI) approaches [1, 2, 3, 4, 5, 6, 7, 8]. The early detection of faults may help to avoid system breakdowns and material damages. A fault can be generally defined as an unpermitted deviation of at least one characteristic property or parameter of a system from a normal condition, e.g., a sensor malfunction. All an unexpected variations that tend to degrade the overall performance of the system can be also interpreted as faults. Contrary to the term failure, which suggests a complete breakdown of the system, the notion fault is used to denote a malfunction rather than a catastrophe. Indeed, failure can be defined as a permanent interruption of the system ability to perform a required function under specified operating conditions. Thus, the role of the fault diagnosis is to monitor the behaviour of the system and to provide all possible information regarding its abnormal functioning.

The most widely known fault diagnosis approach is based on a model of a diagnosed system. Model-based fault diagnosis can be defined as a detection, isolation and identification of faults in the system based on a comparison of system available measurements, with information represented by system mathematical model [1]. Such defined task can be realized in three following steps [9]:

- Residual generation – generation of signals that reflect the fault. Typically, the residual is defined as a difference between outputs of the system and its estimate obtained with a mathematical model.
- Residual evaluation – logical decision making on the time of occurrence and location of faults.
- Fault identification – determination of the type of a fault, its size and cause.

M. Mrugalski, *Advanced Neural Network-Based Computational Schemes for Robust Fault Diagnosis*, Studies in Computational Intelligence 510,
DOI: 10.1007/978-3-319-01547-7_1, © Springer International Publishing Switzerland 2014

The fault detection is the most important stage of the fault diagnosis process. It follows from the fact that without correctly done fault detection it is impossible to perform fault isolation and identification. The main problem that has to be overcome while designing fault detection schemes is that most of them suffer from model uncertainty, i.e., there is a mismatch between the model and the system being considered [10] what results in the undetection of the faults or appearing of false alarms. Among numerous model-based fault detection methods the parameters estimation, observer-based and parity relation are the most often applied [1, 3, 7, 11, 12, 13, 14]. These methods are based on the analytical models created on the physical laws describing system behavior.

Unfortunately, the fault detection systems based on the parameters estimation [7, 13] are only used in the case when the parameters of the diagnosed system have physical meaning. Another factor which limits their scope of the application is that they can be effectively applied only for models which are linear to parameters. It follows from the fact that there is a lack of efficient parameters estimation techniques for a non-linear systems. On the other hand, the problem of robustness can be relatively easily solved by the application of various robust parameters estimation methods [15, 16].

The parity relation methods [1, 11, 17] can be directly applied only for the diagnosis of the linear systems. The problem of the robustness is solved by the introduction of an unknown input representing the measurement noises, disturbances and model uncertainty. In fact the fault detection system is designed in such a way to eliminate the influence of the unknown input on the residual signal and hence it is robust to uncertainty. The known parity relation methods for the non-linear systems [18] require very restrictive conditions according to a form of the model of the diagnosed system, what limit the scope of their application.

The robust observers or filters in the stochastic case are one of the most often applied in the literature solution to the problem of robust fault detection [1, 6, 11, 19, 20]. Such approaches can tolerate a degree of model uncertainty, and hence increase the reliability of fault diagnosis. Similarly, in the case of parity relation methods, the model-reality mismatch is represented by an unknown input and hence a state estimate, and consequently an output estimate, is obtained by taking into account model uncertainty. It means that the effect of the unknown input on the both state estimation error and the residual is minimized. Logically, the number of real applications not only simulated examples should proliferate, yet this is not the case. It seems that there are two main reasons why strong formal methods are not accepted in the engineering practice. Firstly, the design complexity of most observers for the non-linear systems does not encourage engineers to apply them in practice. Secondly, the application of observers is limited by the need for the non-linear state-space models of the system being considered, which is usually a serious problem in complex industrial systems. It explains why most of the examples considered in the literature are devoted to simulated or laboratory systems, e.g., known two- or three- tank system, the inverted pendulum, etc. [3]. Another difficulty is that there are examples for which faults direction is very similar to that of the unknown input. It may lead to a situation in which the effect of some faults is minimized and hence they may be impossible to detect. Other approaches that make use of the idea of the unknown input also inherit these drawbacks, e.g., robust parity relations approaches.

An obvious solution to the above-mentioned problems is to describe model uncertainty in a different way. One of the possible approaches is to use statistical techniques [16, 21] to obtain parametric uncertainty of the model and, consequently, model output uncertainty. This kind of knowledge makes it possible to obtain an adaptive threshold [9, 22] that permits robust fault detection. Unfortunately, the results provided by such techniques are valid only when:

- The structure of the model is the same as that of the system (no structural error).
- The model is linear with respect to parameters. In order to make this approach useful for non-linear-in-parameter models, a chain of restrictive assumptions has to be performed, e.g., linearisation. These assumptions may drastically degrade the reliability of the fault detection procedure.
- The noise nature is assumed to be known, e.g., it is zero-mean gaussian white noise.
- A large number of data is available.

As it was already mentioned the above presented fault diagnosis methods are based on the analytical models which are created on the physical laws describing the system behavior [1, 3, 7, 11]. Unfortunately, in the case of some contemporary industrial systems it is often not possible to create such models because these laws are too complex or unknown. In order to solve this problem a system identification can be applied [23, 24]. The system identification is a two stages process, which relies on a model structure selection and parameter estimation. It is crucial to use such identification technique as it enables reduction of the contribution of the structure errors and the parameter estimates inaccuracy to the model uncertainty.

To overcome these problems and to achieve robust fault detection scheme for a non-linear systems, the system identification approaches which are based on the application of the Artificial Neural Networks (ANNs) [25, 26, 27, 28, 29, 30, 31, 32, 33, 34]. The ANNs can be most adequately characterized as computational models with particular properties such as the ability to learn, adapt and parallel data processing. Other advantages of the ANNs over conventional identification methods include the simplicity of the implementation, generalization abilities and good approximation of non-linear systems. Unfortunately, the ANNs have also some important disadvantages, which limit the effectiveness of the developed FDI systems. The most important disadvantages are inefficient quality of the neural model, not mature approaches allowing modeling of dynamics, and usually not available description of a neural model in the state-space representation. Among other disadvantages it is worth to emphasis that only rare approaches ensure the stability of the neural models during the process of the dynamic system identification. Moreover, there is a limited number of approaches that allow mathematically to describe the neural model uncertainty and this factor has the main impact on the effectiveness of the FDI systems.

In the case of the fault diagnosis the neural models are mostly applied to the generation of the residuals signals. The fault detection process usually relies on the application of a constant threshold. In this approach it is assumed that the residual, which is a difference of the system and the nominal model response, is distinguishably different from zero in a faulty case. If this condition is fulfilled, the faults are detected when the residual crosses the arbitrary defined constant threshold. The examples of using such an approach with the classic Multi-Layer Perceptron (MLP) are leakages detection in an electro-hydraulic cylinder drive in a fluid power system [35], the diagnosis of non-catastrophic faults in a nuclear plant [36], and process valve actuator fault diagnosis [37]. Similar examples, but with dynamic neural networks, are the diagnosis of a chemical plant [38], a monitoring and control of foundry processes [39], the diagnosis of a valve actuator [40], and a steam generator [41, 42].

The weakness of the fault detection method based on the constant threshold relies on the lack of robustness what results in an inappropriate work of the fault detection system e.g., the undetected faults or false alarms appear. The robustness is especially important in practical implementations where various sources of uncertainty may be present, e.g., differences between various copies of a given component, time-varying properties, measurements noise, external disturbances and especially the model uncertainty consisting of structure errors and the parameter estimation inaccuracy. Thus, the effectiveness of the neural model-based FDI approaches preponderantly depends on the quality of the neural model obtained during the system identification which is usually uncertain. The inappropriate selected structure of the neural network is the main factor which influences the quality of the neural model. In order to improve the quality of the neural model the various selection methods of the neural models structure can be applied [43, 44, 45, 46, 47, 48, 49]. The application of the Group Method of Data Handling (GMDH) approach seems to be especially attractive because together with the application of the bounded error set estimation class of algorithms [15, 16], parameter estimates inaccuracy resulting from the measurement uncertainty (i.e. noise, disturbances, etc.) can be reduced.

Irrespective of the identification method used, there is always a problem of model uncertainty. Even though the application of the GMDH approach to the neural model structure selection can improve the quality of the model, the resulting structure is not the same as that of the system. In order to solve such a problem it is necessary to design robust fault detection system.

To achieve this goal it is necessary to mathematically describe the uncertainty of the neural model. This knowledge allows to design the robust fault detection system. The concept of this approach is based on the application of appropriate estimation methods to training of the neural model and obtaining parameters uncertainty. The knowledge about parameters uncertainty makes it possible to calculate the whole neural model uncertainty and in the consequence enables to design so-called adaptive thresholds [9, 22]. The adaptive threshold, contrary to the constant one, bounds the residual at a level that is dependent on the model uncertainty, and hence it provides a more reliable fault detection.

The next desirable future of the neural model is a state-space description. Such representation enables to apply the neural models in the FDI schemes easily. Moreover, the state-space representation simplifies the process of training of the neural models because several efficient parameters estimation algorithms such as the Unscented Kalman Filter (UKF) can be easily applied. Additionally, such representation together with the application of the constrained parameters estimation approach allows to ensure the stability of the neural models during the process of the dynamic system identification.

As it was mentioned only the correctly performed fault detection allows to perform fault isolation and identification. Among numerous fault isolation and identification methods the approaches that use neural networks as pattern classifiers can be found [50, 51, 52, 53]. In such approaches the networks are trained to recognise different modes of the system, i.e. both faulty and non-faulty ones. The examples of using such an approach are the FDI in machine dynamics and vibration problems [54], fault diagnosis of chemical processes [55], gas turbine [42] and induction motor [56, 57]. However, the most popular FDI approach is based on the application of the bank of residual generators [58, 59], which allows to generate the residuals for each fault in the diagnosed system. In the case of this scheme, residual generators are designed in such a way that each residual is sensitive to one fault only while it remains insensitive to others. It should be underlined that in many practical applications such an approach is difficult to realize because it is often not possible to provide the data allowing to design neural models of particular faults. Irrespective of the above difficulties, the above mentioned fault isolation strategy is frequently used in the neural network-based FDI schemes.

In order to solve the challenging problem of the FDI, a combination of the state-space neural models and robust filters can be applied. Indeed, the state-space description of the neural model allows to use the Unknown Input Filter (UIF) and Robust Unknown Input Filter (RUIF) [14, 22, 60] to estimate the inputs and calculate the adaptive thresholds for each input signal of the diagnosed system. In the consequence this approach enables to perform robust FDI of the actuators of the dynamic non-linear systems simultaneously [61, 62]. Thus, the main objective of this book is to present the neural model based FDI methods for non-linear dynamic systems.

1.2 Content

The remaining part of this book is divided into the following main chapters:

Designing of dynamic neural networks. Chapter 2 presents the most popular architectures of dynamic neural networks which can be applied for the identification of non-linear dynamic systems. Moreover, it contains a review of methods which are the most often used for designing of architectures of neural networks. In particular, three main groups of methods are discussed, i.e. Pruning methods, Bottom-up methods and Discrete optimisation methods. Finally, the main advantages and drawbacks of the presented methods are portrayed. The subsequent part of the chapter describes the GMDH approach which can be applied for improvement of neural models quality. In particular, the original developments regarding the dynamic neurons architectures are proposed. The final part of the chapter is devoted to an analysis of sources of the GMDH neural models uncertainty.

Estimation methods in training of ANNs for robust fault diagnosis. Chapter 3 is devoted to parameters estimation methods, which can be applied directly for robust fault diagnosis or can be used for training and uncertainty estimation of the neural networks. On the beginning the concept of the robust FDI via parameters estimation, is presented. In the next part of the chapter a review of parameters estimation methods, which can be used for the parameters uncertainty estimation is presented. In particular, the Least-Mean Square (LMS), Bounded-Error Approach (BEA), Outer Bounding Ellipsoid (OBE) and Zonotope-Based Algorithm (ZBA) are presented. Moreover, the main advantages and drawbacks of the presented techniques are discussed. The remaining part of this chapter shows the effectiveness of the presented parameters estimation techniques on the illustrative example and demonstrates how to employ such approach for the robust FDI of a brushed DC motor.

Neural networks-based robust fault detection of static non-linear systems. Chapter 4 shows original developments regarding neural networks based robust fault detection of a non-linear systems. In particular, the general concept of the robust fault detection with the application of the output adaptive thresholds is proposed. One main objective of this chapter is to show how to describe modelling uncertainty of the MLP and use the resulting knowledge about model uncertainty for robust fault detection. The last part of this chapter presents how to apply such an approach for the robust fault detection of an valve actuator in the Lublin sugar factory.

Robust fault detection of dynamic non-linear systems via GMDH neural networks. Chapter 5 discusses original developments regarding robust fault detection of dynamic non-linear systems with the application of the GMDH neural networks. The original methodologies of the confidence of the GMDH neural models with the application of parameters estimation methods presented in Chap. 3 are proposed. In particular, the methods of calculation of each neuron and the whole GMDH neural network output uncertainty are presented. It should be noticed that the application of the GMDH method decreases the neural model uncertainty and increases the sensitivity of the robust fault detection scheme. In order to show the effectiveness of the GMDH neural models in the robust fault detection tasks, similarly as in Chap. 4, an valve actuator in the Lublin sugar factory is used.

State-space GMDH neural networks for actuator robust fault detection and isolation. Chapter 6 presents original solutions which can be used for the robust FDI of actuator faults. The primary goal of this chapter is to develop a new methodology of designing of a non-linear state-space GMDH neural model, which can be used for the robust fault detection of the system and sensors. This goal is achieved by the application of the UKF to parameters and uncertainty estimation of the state-space GMDH neural model. It should be underlined that the UKF-based constrained parameter estimation algorithm warranties the asymptotically stable GMDH neural model. Moreover, a new and more general GMDH algorithm for a Multi-Input and Multi-Output (MIMO) state-space GMDH neural model is proposed. The secondary goal of this chapter is to develop an original method of the actuator FDI. For this reason two new methods on the basis of application of the UIF and RUIF are proposed. In particular, it is shown how to estimate the GMDH neural model inputs and calculate the input adaptive thresholds. The proposed methods allow to perform the robust FDI of the actuators faults simultaneously. The final part of this chapter is devoted to the application of the proposed approach to the robust FDI of the tunnel furnace.

2

Designing of Dynamic Neural Networks

2.1 Introduction

The most crucial challenge which should be undertaken by engineers who supervise various technical systems is to design a mathematical model describing the system behaviour. The scope of applications of mathematical models in the contemporary industrial systems is extremely broad and includes the FDI [1, 3, 4, 5, 6] and Fault-Tolerant Control (FTC) [63, 64, 65, 66, 67] among others. In the case of a non-linear dynamic system identification the ANNs are often applied [25, 26, 28, 29, 31, 52, 68]. The main advantage of the ANNs is the possibility of obtaining the model which describes the behaviour of the identified non-linear dynamic system only on the basis of measurement data when physical laws describing the system behaviour are not available or they are too complex [11, 63]. It follows from the fact that the ANNs are perceived as universal approximators [25] which can approximate any smooth function with an arbitrary degree of accuracy as the number of hidden layer neurons increases. Other fundamental advantages of neural networks are their learning, generalization and adaptation abilities.

One of the most desirable features of the model obtained during the system identification is its small modeling uncertainty. The system identification is a two stages process, where a model structure selection and parameters estimation is done. It is crucial to use such an identification technique as it enables the reduction of the contribution of the structure errors and the parameters estimates inaccuracy to the model uncertainty. For this reason several methods of designing of neural model architectures were developed. These approaches can be divided into three following classes of methods: Bottom-up, Pruning and Discrete optimisation methods [48]. Unfortunately, the efficiency of those algorithms is usually very limited and usually neural networks with very poor generalisation abilities are obtained.

M. Mrugalski, *Advanced Neural Network-Based Computational Schemes for Robust Fault Diagnosis*, Studies in Computational Intelligence 510,
DOI: 10.1007/978-3-319-01547-7_2,

To overcome this problem the GMDH neural networks [69, 70, 71] have been proposed. The synthesis process of the GMDH neural network is based on the iterative processing of a sequence of operations. This process leads to the evolution of the resulting model structure in such a way so as to obtain the best quality approximation of the identified system. Thus, the task of designing a neural network is defined in order to obtain a model with small uncertainty. The GMDH neural model is perceived as the set of hierarchically connected partial models where each of them should be the best approximate of the identified system. Thus, each partial model should have abilities to model dynamics. For such a reason, a few new structures of dynamic neurons are proposed in this chapter. The application of the classic GMDH approach can improve the quality of the neural model, however, it does not eliminate the whole neural model uncertainty. Still some neural model structure errors or/and parameters estimate inaccuracy can be introduced to the network. The knowledge about potential sources of model uncertainty allows to improve the GMDH approach and adopt the neural model to the application in the robust fault detection systems.

The chapter is organized as follows: Section 2.2 presents the most popular types of dynamic neural networks which can be applied for the dynamic non-linear systems identification. Section 2.3 is devoted to methods for designing of neural model architectures. Section 2.4 discusses the details of synthesis process of the GMDH neural models. Section 2.5 is devoted to the developed structures of the dynamic neurons. Finally, Sect. 2.6 presents the sources of the GMDH neural model uncertainty. In particular, the contribution of the structure errors to the model uncertainty is described in detail. It should be also pointed out that the results described in this chapter are based on [72, 73, 74, 75, 76, 77].

2.2 Dynamic Neural Networks

The ANNs are often applied for the identification of non-linear dynamic systems. The most often used architecture is a feed-forward multilayer network which consists of neurons with monotonic activation functions. This architecture, called the MLP [25, 26, 29, 78], is often modified by the introduction of different feedbacks resulting in the appearance of numerous recurrent neural networks [26, 79, 80, 81, 82]. The MLP is constructed with a certain number of neurons which can have different architecture. Usually, a neuron model is described by the following equation:

$$\hat{y}_k = f\left(\sum_{i=1}^{n_p} \hat{p}_i r_{i,k} + r_0\right) = \hat{\boldsymbol{p}}^T \boldsymbol{r}_k, \tag{2.1}$$

where $\boldsymbol{r}_k \in \mathbb{R}^{n_p}$ is the k-th input vector of neuron, $\hat{\boldsymbol{p}} \in \mathbb{R}^{n_p}$ is the parameters estimate vector, $\hat{y}_k \in \mathbb{R}$ is the neural model output and $f(\cdot)$ is the neuron

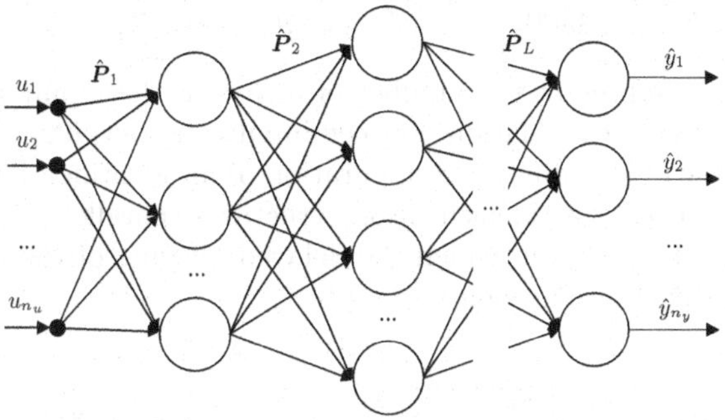

Fig. 2.1. Multi-Layer Perceptron

activation function. The structure of the MLP consists of neurons which are grouped into layers (cf. Fig. 2.1). Such a network has an input layer, one or more hidden layers and an output layer of neurons. The main task of the input layer is preliminary input data processing relying on e.g., scaling, filtering or signal normalization. The main data processing is carried out in the hidden and output layers. It is necessary to notice that connections between neurons are designed in such a way that each element of the previous layer is connected with each element of the next layer. These connections are assigned to suitable parameters so-called weight coefficients which are estimated during network training. The outputs of neural model can be described by the following relation:

$$\hat{\boldsymbol{y}}_k = f_{n_l}(\hat{\boldsymbol{P}}_{n_l}, \ldots, f_2(\hat{\boldsymbol{P}}_2 f_1(\hat{\boldsymbol{P}}_1 \boldsymbol{u}_k))), \tag{2.2}$$

where f_1, f_2 and f_{n_l} denote the non-linear operators which define a neural signal transformation through the subsequent layer of network, $\hat{\boldsymbol{P}}_1$, $\hat{\boldsymbol{P}}_2$ and $\hat{\boldsymbol{P}}_{n_l}$ represent the matrices of parameters, $\boldsymbol{u}_k \in \mathbb{R}^{n_u}$ and $\hat{\boldsymbol{y}} \in \mathbb{R}^{n_y}$ denote the input and output vectors of data and $k = 1, ..., n_{\mathcal{T}}$ where $n_{\mathcal{T}}$ is the number of samples in the training data set. The fundamental advantages of neural networks are learning, adaptation and generalization abilities. From technical point of view, the training of neural network relies on the determination of parameters values between the neurons in order to minimize an early defined goal function. The fundamental training algorithm for feed-forward multi-layer networks is the Back-Propagation (BP) algorithm [3, 26, 78, 83]. This iterative algorithm is based on the minimization of a sum-squared error using the gradient descent method.

Most of the industrial systems are dynamic and non-linear in its nature, and hence during their identification it seems desirable to employ the models which can represent the dynamic of the system. In the case of the static neural network, as the MLP, the modelling problem of dynamic is usually tried to be solved by the introduction of additional, suitably delayed global output feedback lines [26, 84]. Let us assume that a dynamic process is described by the following non-linear discrete difference equation:

$$\boldsymbol{y}_k = f(\boldsymbol{u}_k, \boldsymbol{u}_{k-1}, \ldots, \boldsymbol{u}_{k-n_u}, \boldsymbol{y}_{k-1}, \ldots, \boldsymbol{y}_{k-n_y}), \tag{2.3}$$

where n_u and n_y represent the number of delays in the inputs and outputs and $f(\cdot)$ is the non-linear relation between inputs and outputs describing the system behaviour. According to configuration of an identification scheme, it is possible to specify two identification structures, parallel [85] and series-parallel [86] (cf. Fig. 2.2), in which the delayed outputs of the model $\hat{\boldsymbol{y}}_k$ or system $\boldsymbol{y}_k$ are taken into account.

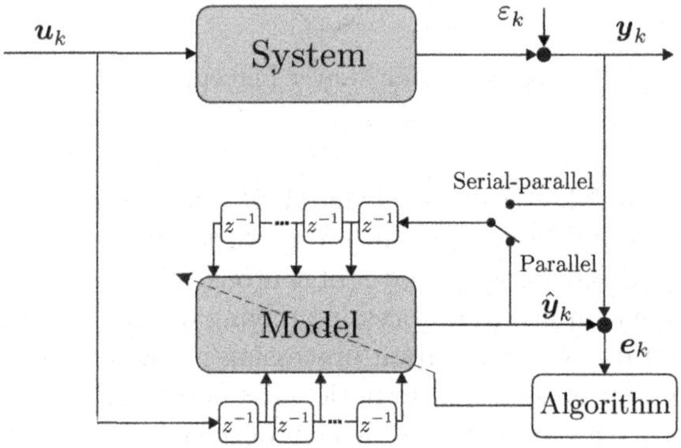

Fig. 2.2. Serial-parallel and parallel structures of the identification

In the case of parallel identification structure:

$$\hat{\boldsymbol{y}}_k = f(\boldsymbol{u}_k, \boldsymbol{u}_{k-1}, \ldots, \boldsymbol{u}_{k-n_u}, \hat{\boldsymbol{y}}_{k-1}, \ldots, \hat{\boldsymbol{y}}_{k-n_y}), \tag{2.4}$$

it is difficult to assure the stability of the model obtained during identification. In other words, even though signals $\boldsymbol{u}_k$ and $\boldsymbol{y}_k$ are limited, the identification error $\boldsymbol{e}_k$ must not be convergent on zero. From this reason, during the system identification much more often is applied the series-parallel structure.

$$\hat{\boldsymbol{y}}_k = f(\boldsymbol{u}_k, \boldsymbol{u}_{k-1}, \ldots, \boldsymbol{u}_{k-n_u}, \boldsymbol{y}_{k-1}, \ldots, \boldsymbol{y}_{k-n_y}), \tag{2.5}$$

Unfortunately, this kind of structure cannot be used in the FDI or FTC tasks, where the model is used to generate the residual signal. If it is assumed that the identification error converge to zero, then output signals of the model and system can be described by the following relation:

$$\boldsymbol{y}_k \approx \hat{\boldsymbol{y}}_k. \tag{2.6}$$

From this reason, it is possible to use the parallel architecture during the final stage of the identification process and the application in the fault diagnosis task. In this way, the neural network described by (2.4) uses its own outputs as a part of the input space. Thus, feedback from the network output to its input is introduced, and the feed-forward network becomes a recurrent network with an outer feedback [87, 88].

The classical feed-forward networks like the MLP can be only used for the identification of the static non-linear systems. Therefore, in the case of dynamic system identification with this kind of the network the approach based on the feedback from inputs and outputs has to be applied. Unfortunately, such methodology has some drawbacks e.g., the number of delays in the inputs and outputs, which represents the order of dynamic of the identified system, should be known. In numerous cases when the order of dynamic can not be obtained analytically it has to be selected experimentally what extends the identification process. Moreover, as the bigger number of the network inputs exists then in the consequence the larger number of parameters should be estimated. Furthermore, the problems with convergence of learning process and stability of the obtained model can appear when feedback connections in the neural network are introduced.

In order to solve all these problems several fully or partially recurrent neural networks have been developed which are the modification of the MLP relying on the introducing connections between neurons and introducing local feedbacks in neurons. The recurrent networks [86] are characterized by better properties from the point of view of their application to control theory. As a result of introducing feedback into the network structures, it is possible to accumulate information and use it later. Depending on the form of the feedback connection it is possible to distinguish local and global recurrent neural networks [80]. The local recurrent networks have a structure similar to static feed-forward ones but they consist of dynamic neuron models with internal feedback. The details of such neurons will be presented in Sect. 2.5. In the global recurrent networks a feedback is allowed between neurons of different layers or between neurons of the same layer.

The first network belonging to the class of the global recurrent networks is the Real Time Recurrent Network (RTRN) [79]. This network consists of one layer of neurons but only a part of them generates the output signals. All neurons outputs are delayed and introduced as additional inputs to the network which can stimulate each neuron (cf. Fig. 2.3). Thus, any connections between the neurons are allowed and a fully connected neural architecture is obtained. This relatively simple network architecture is designed for real time signals processing and this is a reason why this network is called the RTRN. The fundamental advantage of such networks is the possibility of approximating of a wide class of dynamic systems. Unfortunately, training of the RTRN network is usually complex and slowly convergent [87]. Moreover, there are problems with keeping network stability.

The Elman neural network [89] also belongs to the class of global recurrent networks. This network consists of four layers of neurons: the input, context, hidden and output layers (cf. 2.4). The neurons in input and output layers interact with the outside environment, whereas the hidden and context units do not. The neurons in the context layer are only used to memorize the

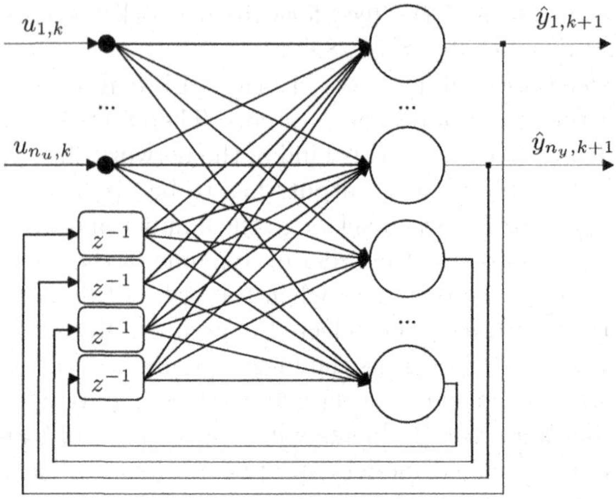

Fig. 2.3. Real Time Recurrent Network

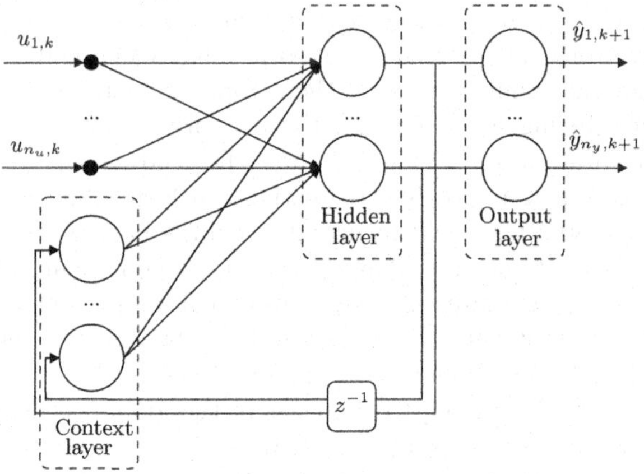

Fig. 2.4. Elman neural network

previous activations of the hidden neurons. In the Elman network the number of context neurons is equal to the number of neurons in the hidden layer. All feed-forward connections are adjustable and the recurrent connections are fixed. The assumed number of neurons in the context layer represents the order of dynamics of the identified system [87]. The behaviour of the Elman network is similar to a feed-forward network. Therefore, the standard BP algorithm can be applied to estimate the network parameters.

Next global recurrent network created on the basis of the feed-forward network is the Recurrent Multi-Layer Perceptron [90]. In this network the local feedbacks in neurons and feedbacks between neurons in the layer exist (cf. Fig. 2.5). In this way the feed-forward connections are responsible for

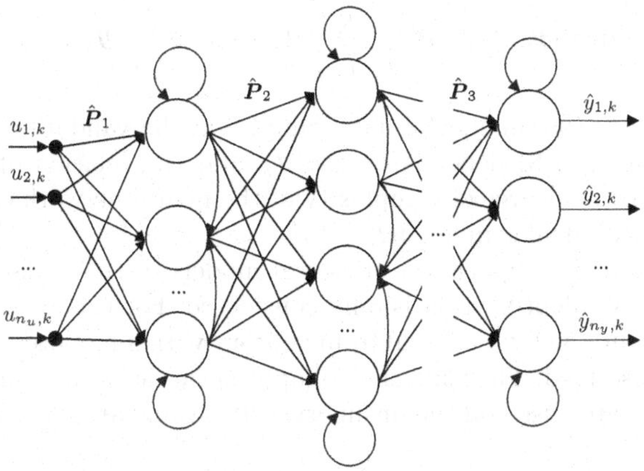

Fig. 2.5. Recurrent Multi-Layer Perceptron

modelling of non-linearities and feedbacks for dynamics of the identified system. The recurrent MLP network has been successfully used in dynamic system identification tasks [90]. However, a drawback of this dynamic structure is an increased network complexity and resulting longer training time.

Unfortunately, all global recurrent neural networks, in spite of their usefulness in control theory, have some disadvantages. These architectures suffer from a lack of stability and complicated training process. Moreover, for such neural models a mechanism of automatical structure selection does not exist what may result in the increase of model uncertainty. Another disadvantage of the global recurrent networks is a lack of state-space description what is a crucial future in the case of its application in the FDI and FTC systems.

2.3 Methods for Designing of Neural Models

The system identification with the application of the ANNs relies on the choice of the neural model architecture N_{arch} from the space of all possible networks architectures $\mathbb{N}_{\text{arch}}$ and its parameters estimates $\hat{\boldsymbol{p}}$ on the basis of training data set $\mathcal{T}$ in order to minimize the following cost function:

$$J_{\mathcal{T}}(\boldsymbol{y}, \hat{\boldsymbol{y}}(N_{\text{arch}}, \hat{\boldsymbol{p}})), \tag{2.7}$$

where $\hat{\boldsymbol{y}}(N_{\text{arch}}, \hat{\boldsymbol{p}})$ denotes the output of the neural model and $\boldsymbol{y}$ represents the system output. One of the most important features of the neural models are generalization abilities [91] which are defined as the possibility of generation of appropriate output signals for the validation data set $\mathcal{V}$ not used during parameters estimation. The generalization abilities of the neural model can be described by the following quality index $J_{\mathcal{V}}$:

$$J_{\mathcal{V}}(\boldsymbol{y},\hat{\boldsymbol{y}}(N_{\text{arch}},\hat{\boldsymbol{p}})) = \sum_{k=1}^{n_{\mathcal{V}}} (\hat{\boldsymbol{y}}_k(N_{\text{arch}},\hat{\boldsymbol{p}}) - \boldsymbol{y}_k)^2, \tag{2.8}$$

where $n_{\mathcal{V}}$ denotes the number of data samples in the validation data set $\mathcal{V}$. The quality index (2.8) depends on the $J_{\mathcal{T}}$ and the relation of number of data samples to the degree of complexity of the neural network expressed by the Vapnik-Chervonenkis index [92].

In this section the application of neural models in the tasks of dynamic non-linear system identification is only considered. For this reason, only the feed-forward neural networks build from neurons with a monotonic activation function are described. The wide class of approaches for self-organized neural networks which are based on the unsupervised learning are not presented [26, 93, 94].

2.3.1 Pruning Methods

Pruning methods gradually decrease the structure of the neural model which is potentially sufficient to the modelling of the relation between inputs and outputs of the identified system. The reduction of the neural model is done by gradually decreasing parameters of neurons what results in removing of neurons from the network. Such procedure increases the generalisation properties of the neural model. Among pruning methods the Sensitivity Methods, Penalty Function Methods and methods based on the Information Analysis can be distinguished [48, 49].

In the Sensitivity Methods the completed structure of the neural model is chosen and preliminary parameters estimates are calculated. In the next step the values of the sensitivity measures $s_{i,j}$, describing the influence of neuron parameters on function (2.8) are obtained. The parameters of the neurons for which the values of $s_{i,j}$ are small are removed from the network. Such process is repeated until all parameters with the smallest influence on $J_{\mathcal{V}}$ are removed. In the literature numerous definitions of the sensitivity measures can be found. One of them is proposed in paper [43]:

$$s_{i,j}^{MS} = -\frac{\partial J_{\mathcal{V}}}{\partial \alpha_{i,j}}\Big|_{\alpha_{i,j}=1}, \tag{2.9}$$

where $\alpha_{i,j}$ denotes additional set of coefficients connected with j-th parameter of i-th neuron $\hat{y}_i = f(\sum_j p_{i,j}\alpha_{i,j}u_j)$. If $\alpha_{i,j} = 0$ then the present connection is removed from the network, opposite to the case when $\alpha_{i,j} = 1$.

One of the most popular sensitive measure is proposed in the Optimal Brain Damage (OBD) method [95, 96]. This approach is based on the expansion of the function $J_{\mathcal{V}}(\hat{\boldsymbol{p}})$ into Taylor series around the current solution $\hat{\boldsymbol{p}}$:

$$J_{\mathcal{V}}(\boldsymbol{p})-J_{\mathcal{V}}(\hat{\boldsymbol{p}}) = \nabla^T J_{\mathcal{V}}(\hat{\boldsymbol{p}})(\boldsymbol{p}-\hat{\boldsymbol{p}})+\frac{1}{2}(\boldsymbol{p}-\hat{\boldsymbol{p}})^T \times \boldsymbol{H}(\hat{\boldsymbol{p}})(\boldsymbol{p}-\hat{\boldsymbol{p}})+o(\boldsymbol{p}-\hat{\boldsymbol{p}}). \tag{2.10}$$

In the OBD method it is assumed that the value $o(\boldsymbol{p}-\hat{\boldsymbol{p}})$ is negligibly small and $\nabla^T J_{\mathcal{V}}(\hat{\boldsymbol{p}}) = 0$ because the reduction of parameters is performed after the training, when $J_{\mathcal{V}}(\hat{\boldsymbol{p}})$ receives the minimum value for $\hat{\boldsymbol{p}}$. Additionally, in order to simplify the OBD method it is assumed that the diagonal elements of the Hessian matrix $\boldsymbol{H}$ are predominant what allows to consider only the diagonal elements $h_{k,k} = \frac{\partial J_{\mathcal{V}}}{\partial \hat{p}_{i,j}^2}$, what results in the following sensitivity measure:

$$s_{i,j}^{\mathrm{OBD}} = \frac{1}{2}\frac{\partial J_{\mathcal{V}}}{\partial \hat{p}_{i,j}^2}\hat{p}_{i,j}^2, \tag{2.11}$$

where $\hat{p}_{i,j}$ denotes j-th connection of i-th neuron. The application of the OBD method allows to obtain the neural network with good generalization properties. Moreover, in a single step numerous connections can be pruned. The main disadvantage of such method relies on repeating of learning process after the pruning of the network structure.

The improved version of the OBD algorithm is the Optimal Brain Surgeon (OBS) method [46]. In this approach, also the expansion of the function $J_{\mathcal{V}}(\hat{\boldsymbol{p}})$ into Taylor series around the current solution $\hat{\boldsymbol{p}}$ is performed and the first order factor is removed. However, all components of the Hessian matrix are taken into account:

$$s_i^{\mathrm{OBS}} = \frac{1}{2}\frac{\hat{p}_i^2}{[\boldsymbol{H}^{-1}]_{ii}}. \tag{2.12}$$

The main advantage of the OBS method is simple sensitivity measure allowing for correcting of not pruning parameters which minimizes the goal function in spite of pruning insignificant connection:

$$\delta\hat{\boldsymbol{p}} = \frac{\hat{p}_i^2}{[\boldsymbol{H}^{-1}]_{ii}}\boldsymbol{H}^{-1}\boldsymbol{e}_i, \tag{2.13}$$

where $\boldsymbol{e}_i$ represents unitary vector with ones at i-th position. The disadvantage of the OBS method is higher complexity following from the calculation of all elements of the Hessian matrix. Moreover, in the OBS method only single connection can be pruned at each step of the algorithm.

The next group of approaches belonging to pruning methods are the Penalty Function Methods. The concept of such methods relies on modification of the goal function $J_{\mathcal{V}}$ by adding a penalty factor $\Gamma(\hat{\boldsymbol{p}})$ for redundant connections or redundant neurons:

$$J'_{\mathcal{V}}(\boldsymbol{y}, \hat{\boldsymbol{y}}(N_{\mathrm{arch}}, \hat{\boldsymbol{p}})) = J_{\mathcal{V}}(\boldsymbol{y}, \hat{\boldsymbol{y}}(N_{\mathrm{arch}}, \hat{\boldsymbol{p}})) + \gamma\Gamma(\hat{\boldsymbol{p}}), \tag{2.14}$$

The penalty factor $\Gamma(\hat{\boldsymbol{p}})$ for redundant connections causes the minimization of the neurons parameters during the network training. On the beginning the cost function $J_{\mathcal{V}}(\cdot)$ is minimized during the estimation of the parameters $\hat{\boldsymbol{p}}_n$. In the next stage the estimated parameters are corrected according to the following relation:

$$\hat{\boldsymbol{p}} = \hat{\boldsymbol{p}}_n - \hat{\boldsymbol{p}}_n\eta\Gamma_K(\hat{\boldsymbol{p}}_n), \tag{2.15}$$

where η represents learning constant. In the literature several penalty functions and correction functions of parameters can be found [97, 98, 99]:

$$\Gamma(\hat{\boldsymbol{p}}) = \|\hat{\boldsymbol{p}}\|^2, \quad \Gamma_K(\hat{p}'_{i,j}) = \gamma. \tag{2.16}$$

The disadvantage of the penalty function (2.16) is an equal correction of all parameters. In order to solve this problem the following penalty function can be applied:

$$\Gamma(\hat{\boldsymbol{p}}) = \frac{1}{2}\sum_{i,j}\frac{\hat{p}_{i,j}^2}{1+\hat{p}_{i,j}^2}, \quad \Gamma_K(\hat{p}'_{i,j}) = \frac{\gamma}{(1+\hat{p}'^2_{i,j})^2}. \tag{2.17}$$

It should be underlined that for small values of parameters ($\hat{p}_{i,j} \ll 1$) penalty function (2.16) and (2.17) are approximately equal, although when ($\hat{p}_{i,j} \gg 1$) the correction of parameters is small. The generalization of (2.17) is the following penalty function:

$$\Gamma(\hat{\boldsymbol{p}}) = \frac{1}{2}\sum_{i,j}\frac{\hat{p}_{i,j}^2}{1+\sum_k \hat{p}_{i,k}^2}, \quad \Gamma_K(p'_{i,j}) = \frac{\gamma(1+2\sum_{k\neq j}\hat{p}'^2_{i,j})}{(1+\sum_k \hat{p}'^2_{i,k})^2}. \tag{2.18}$$

The minimization of (2.18) leads to pruning neurons parameters and some neurons simultaneously, for which the norm of the parameters vector fulfills the following relation:

$$\sum_k \mid \hat{p}_{i,k} \mid \approx 0. \tag{2.19}$$

The next class of penalty function methods relies on the introduction of the penalty factor $\Gamma(\hat{\boldsymbol{p}})$ for redundant neurons. In such an approach it is assumed that the neuron with a small output activity for a significant change of input signals is redundant because it does not play an important role in signal processing. On the basis of this assumption in the work [100] the penalty function was modified by adding the following penalty factor:

$$\Gamma(\hat{\boldsymbol{p}}) = \mu\sum_i\sum_j e(\Delta_{i,j}^2), \tag{2.20}$$

where the sum is calculated on the basis of all j learning samples and i hidden neurons and parameter μ represents the influence of the correction factor on the goal function. The correction element $e(\Delta_{i,j}^2)$ is assumed in such a way that it ensures large corrections of neurons parameters for small changes of $\Delta_{i,j}$ and conversely ensure small corrections of neurons parameters for significant changes of $\Delta_{i,j}$:

$$e = \frac{\partial e(\Delta_{i,j}^2)}{\Delta_{i,j}^2} = \frac{1}{(1+\Delta_{i,j}^2)^n}. \tag{2.21}$$

The parameter n allows to modify the form of the penalty function. If it is assumed that $n = 0$ then (2.21) receives the form of the linear correction function $e = \Delta_{i,j}^2$ what denotes equal punish of all neurons. When it is assumed that $n = 1$ then the logarithmic correction function $e = \ln(1+\Delta_{i,j}^2)$ is assumed leading to significant punish of neurons with small activity.

In the methods based on Information Analysis [101] the neural model structure optimisation is performed by the analysis of the eigenvalues of the matrix $\boldsymbol{C}$ which is associated with the outputs of neurons from the hidden layer. This matrix is obtained on the basis of the $n_{\mathcal{T}}$ learning samples:

$$\boldsymbol{C} = \left[c_{i,j} = \frac{1}{n_{\mathcal{T}}} \sum_{k=1}^{n_{\mathcal{T}}} (\hat{y}_{i,k} - \bar{y}_i)(\hat{y}_{j,k} - \bar{y}_j)\right], \tag{2.22}$$

where $\hat{y}_{i,k}$ denotes output of i-th hidden neuron obtained for k-th input sample and $\bar{y}_i = \frac{1}{n_{\mathcal{T}}} \sum_{k=1}^{n_{\mathcal{T}}} (\hat{y}_{i,k})$. The covariance matrix is symmetric and positive semi-defined [101], thus it can be transformed to a diagonal form using some orthonormal matrix $\boldsymbol{V}$:

$$\boldsymbol{C} = \boldsymbol{V} \mathrm{diag}(\lambda_i | i = 1, \ldots, n_y) \boldsymbol{V}^T, \tag{2.23}$$

where n_y denotes the number of neurons in the current hidden layer. According to the methods based on information analysis neurons corresponding negligible low eigenvalues λ_i are removed from the hidden layer [101]. Unfortunately, the pruned networks have to be retrained what results in the extension of the identification procedure. This weakness is deprived by the solution proposed in [102] where for the neural network, which is consisted of one hidden layer of neurons and one output neuron, an additional virtual layer with the same number of neurons is included. The set of parameters between neurons in the hidden and virtual layers represents the matrix $\boldsymbol{V}$ and parameters between neurons in the hidden layer and output neuron represents the vector $\boldsymbol{V}^T\hat{\boldsymbol{p}}$, respectively. The vector $\hat{\boldsymbol{p}}$ denotes a set of parameters between neurons located in the hidden layer and output neuron when the network of the virtual layer of neurons is not included. The output signal of such network is generated according to the following relation:

$$\hat{\boldsymbol{y}}(N_{\mathrm{arch}}, \hat{\boldsymbol{p}}) = \hat{\boldsymbol{p}}^T \boldsymbol{V} (\boldsymbol{V}^T \boldsymbol{u}) = \hat{\boldsymbol{p}}^T \boldsymbol{u}. \tag{2.24}$$

On the basis of (2.23) it can be noticed that the covariance matrix $\boldsymbol{C}_v$ associated with the outputs of the virtual neurons is diagonal:

$$\boldsymbol{C}_v = \mathrm{diag}(\lambda_i | i = 1, \ldots, n_y). \tag{2.25}$$

If it is assumed that the values of the variance λ_i are negligibly small, then it can be assumed that $\lambda_i = 0$ and in the consequence i-th virtual neuron achieves the constant mean value $\bar{y}_i$ regardless of values of input signals. The constant value of the virtual neuron output allows to add this signal to the bias of the output neuron and the virtual neuron can be pruned from the network. The process of removing of virtual neurons begins from neurons associated with the smallest value of covariance matrix and is repeated until increasing $J_{\mathcal{T}}$ (2.8). The presented approach allows to obtain the optimal network architecture in the sense of generalisation ability, however, the achieved model still has the redundant parameters.

2.3.2 Bottom-Up Methods

The concept of the Bottom-up methods relies on gradually increase of a small neural network until it achieves a sufficient complexity to identify the considered system. Among numerous Bottom-up methods a wide class of methods for the selection of the feed-forward network consists of neurons with a step activation function exists. Between these algorithms the Machands [45], Tiling [47] and Upstart [103] algorithms can be distinguished. Moreover, a limited number of methods for the feed-forward neural networks with a monotonic activation function in neurons can be found. Such methods, opposite to the earlier mentioned, can be applied to the neural model structure selection in the non-linear systems identification tasks.

The Cascade Correlation Method [44] is one of the most widely applied algorithms in the systems identification tasks. The concept of this algorithms is based on the iterative improvement of the neural model quality by the introduction of the hidden neurons to the network. The network synthesis process starts from the determination of the number of input signals and output neurons. In the next stage the learning of the output neurons with the application of the BP algorithm is performed. If the resulting neural model is satisfying the algorithm is stopped otherwise a new neuron (candidate neuron) in the hidden layer is added. The input signals of the added neuron consist of all inputs of the neural network and all outputs of the hidden neurons earlier added to the network. However, before the candidate neuron is connected to the network all its parameters have to be estimated by the BP algorithm according to the following quality criterion:

$$J_{\mathcal{C}} = \sum_{i=1}^{n_y} \left| \sum_{k=1}^{n_{\mathcal{T}}} \left(\hat{y}_{c,k} - \bar{y}_c\right)\left(\varepsilon_{i,k} - \bar{\varepsilon}_i\right) \right|, \tag{2.26}$$

where n_y denotes the number of output neurons, $\hat{y}_{c,k}$ is the response of the candidate neuron, and $\varepsilon_{i,k}$ represents the output error of i-th output neuron. The values $\bar{y}_c$ and $\bar{\varepsilon}_i$ are defined as follows:

$$\bar{y}_c = \frac{1}{n_{\mathcal{T}}} \sum_{k=1}^{n_{\mathcal{T}}} \hat{y}_{c,k}, \tag{2.27}$$

and

$$\bar{\varepsilon}_c = \frac{1}{n_{\mathcal{T}}} \sum_{k=1}^{n_{\mathcal{T}}} \varepsilon_{i,k}. \tag{2.28}$$

After the termination of the training process the parameters of the candidate neuron are frozen and the neuron is connected to the network. At the next step the parameters of the output neurons are estimated once again. The introduction of new neurons to the network is stopped when the assumed output error of the neural model is achieved. The obtained network is optimal from the quality modeling point of view, however, it is not optimal from the number of neurons and their parameters point of view.

The Dynamic Node Correction [104] is the algorithm which gradually increases the size of the neural network and it is designed for the synthesis of the perceptron network consisting of one hidden layer with the application of the BP algorithm. The synthesis process begins from the parameters estimation of a small group of neurons. In the next step new neurons are introduced to the network and training process is repeated. It should be emphasised that after the introduction of the subsequent neurons to the neural network, only their parameters are randomly initialized when the remaining parameters are assumed on the basis of parameters obtained in the previous learning process.

The Canonical Form Method is based on the theorem which defines necessary and sufficient conditions on the existing neural approximation of canonical decomposition of a continues function [105]. This approach allows to select the structure of the neural network which consists of two hidden layers of neurons. In this approach the matrix $\boldsymbol{U}$, which columns represent subsequent input learning patterns, is assumed. Moreover, the matrixes $\boldsymbol{Y}_1 = f_1(\boldsymbol{U})$ and $\boldsymbol{Y}_2 = f_2(\boldsymbol{Y}_1)$, which columns respond to the outputs signals from the first and second hidden layer and f_1 and f_2 represent the relations between the the first and the second hidden layer. The optimal number of the neurons n_{y1} in the first and n_{y2} in the second layer has to fulfil the following relations:

$$\det(\boldsymbol{Y}_1^T\boldsymbol{Y}_1)_{n_{y1}} \geq \delta, \quad \det(\boldsymbol{Y}_1^T\boldsymbol{Y}_1)_{n_{y1}+1} < \delta, \tag{2.29}$$

$$\det(\boldsymbol{Y}_2^T\boldsymbol{Y}_2)_{n_{y2}} \geq \delta, \quad \det(\boldsymbol{Y}_2^T\boldsymbol{Y}_2)_{n_{y2}+1} < \delta, \tag{2.30}$$

where $\det(\boldsymbol{Y}_i^T\boldsymbol{Y}_i)_{n_{yi}}$ denotes the matrix $\boldsymbol{Y}_i$ with n_{yi} rows. The process of selection of neurons number in the hidden layers starts from choosing small values of n_{y1} and n_{y2}. In the next step, after the training the verifications of the conditions n_{y1} and n_{y2} are performed. If one of them is not fulfilled then the subsequent hidden neuron to the adequate layer of the network is added and all process is repeated once again.

2.3.3 Discrete Optimisation Methods

In the Discrete Optimisation Methods it is assumed that the ANNs, which are obtained during the training with their quality indexes, can be represented by the graphs, trees or matrixes, which are searched by the desecrate optimisation algorithm. The solution of such problem can be found with the application of genetic [106, 107, 108, 109], evolutionary [110, 111] or $A^\star$ algorithm [112, 113].

The genetic algorithms, on the grounds of the method of representation of information about the architecture of the neural model in the form of the chromosome structure, are especially attractive tools to find an optimal architecture of the neural model. One of the simply representation is a string of bits [114]. In such representation an upper limit of the ANN architecture complexity N_{arch}^{M} is defined. Next, all neurons in each layer of N_{arch}^{M} are numbered from 1 to N. In this way the searching space is limited to a class of digraphs of N nodes. Any architecture of the neural network included in the N_{arch}^{M} is represented by the incidence matrix $\boldsymbol{V}_{\text{inc}}$. Each element equaled to $V_{i,j} = 1$ or $V_{i,j} = 0$ depends on the existing or no connection between i-th and j-th neurons in the network.

On the beginning of the algorithm the initial set of the incidence matrixes is randomly generated. On the basis of the incidence matrixes the chromosomes are created, which are processed by the application of the mutation and crossover operators. Thanks to the reproduction new architectures of neural networks are created which constitute the potential solution of the optimisation a problem. Unfortunately, the convergence of the described approach can be very slow especially for a long genotype corresponding to a large architecture of the neural model. The presented method of the representation of the neural model structure, where each bit of chromosome represents single connection between neurons, is called direct encoding [115]. In the works [106, 107] the indirected encoding method which additionally allows to represent the number of neurons and layers of neurons have been developed. Moreover, the information about parameters of training algorithms can be represented in this approach. In this way the genetic algorithm process searches for the optimal neural model architecture and training process simultaneously.

The $A^{\star}$ algorithm [112] also belongs to the discrete optimisation methods. This algorithm allows to find subset of nodes $\mathcal{G}$ which is included in the set of all architectures of neural networks $\mathcal{N}_{\text{arch}}$. The subset $\mathcal{G}$ represents the optimal architecture of the network for which the training process is finished with the appropriate low value of the quality criterion $J_{\mathcal{V}}$:

$$\mathcal{G} = \left(N_{\text{arch}} \in \mathcal{N}_{\text{arch}} \middle| \max_{\mathcal{V}} J_{\mathcal{V}}(\boldsymbol{y}, \hat{\boldsymbol{y}}(N_{\text{arch}}, \hat{\boldsymbol{p}})) \leq \eta_0\right). \tag{2.31}$$

The concept of the algorithm relies on searching of the optimal solution in the graph consisting of nodes e_i with the associated heuristic functions $h(e_i)$ which define the cost of the best path from the node e_i to the node $e_{\mathcal{G}} \in \mathcal{G}$. The optimal path and cost to the node $e_{\mathcal{G}}$ is unknown. On the basis of the heuristic function $h(e_i)$ and cost $g(e_i)$ the best known path from the start node e_0 to the node e_i, it is possible to obtain the cost function for a given node:

$$f(e_i) = g(e_i) + h(e_i). \tag{2.32}$$

In order to apply the $A^{\star}$ algorithm it is necessary to define the following components [48, 112]:

- A set of goal architectures $\mathcal{G} \subset \mathcal{N}_{\mathrm{arch}}$.
- The expansion operator $\Xi(N_{\mathrm{arch}}) : \mathcal{N}_{\mathrm{arch}} \rightarrow 2^{\mathcal{N}_{\mathrm{arch}}}$ which enables the determination of the set of network architectures being successors of the architecture N_{arch}. The operator $\Xi(N_{\mathrm{arch}})$ can generate successor by adding a new hidden layer of neurons directly before the output layer or by adding a new neuron in the hidden layer.
- The cost function $g(N_{\mathrm{arch}}, N'_{\mathrm{arch}})$ associated with each expansion operation:

$$g(N_{\mathrm{arch}}, N'_{\mathrm{arch}}) = \begin{bmatrix} \vartheta_h(N'_{\mathrm{arch}}) - \vartheta_h(N_{\mathrm{arch}}) \\ \vartheta_l(N'_{\mathrm{arch}}) - \vartheta_l(N_{\mathrm{arch}}) \end{bmatrix}, \tag{2.33}$$

where $\vartheta_h(N_{\mathrm{arch}}) = \sum_{i=1}^{L-1} \mathrm{card}(V_i)$ denotes the number of neurons in $\vartheta_l = (l-1)$ hidden layers.
- The heuristic function $h(N_{\mathrm{arch}})$. If it is assumed that $\hat{\boldsymbol{p}}$ represents a vector of parameters estimates for the architecture N_{arch}, then a heuristic vector associated with the current architecture of the network can be defined in the following form:

$$h(N_{\mathrm{arch}}) = \frac{1}{\sup_{\boldsymbol{u}\in\mathcal{V}} J_{\mathcal{V}}(\boldsymbol{y}, \hat{\boldsymbol{y}}(N_{\mathrm{arch}}, \hat{\boldsymbol{p}}_0))} \cdot \begin{bmatrix} \sup_{\boldsymbol{u}\in\mathcal{V}} J_{\mathcal{V}}(\boldsymbol{y}, \hat{\boldsymbol{y}}(N_{\mathrm{arch}}, \hat{\boldsymbol{p}})) \\ 0 \end{bmatrix}, \tag{2.34}$$

where $\sup_{\boldsymbol{u}\in\mathcal{V}} J_{\mathcal{V}}(\boldsymbol{y}, \hat{\boldsymbol{y}}(N_{\mathrm{arch}}, \hat{\boldsymbol{p}}_0))$ represents a generalisation error for the optimal trained architecture of the network N_{arch_0} for which the algorithm $A^\star$ was initialized.
- The relation of linear order of vectors $\boldsymbol{A}, \boldsymbol{B} \in \mathbb{R}^2$,

$$(\boldsymbol{A} \dot{\leq} \boldsymbol{B}) \Leftrightarrow (a_1 \leq b_1) \vee ((a_1 = b_1) \wedge (a_2 \leq b_2)), \tag{2.35}$$

in order to compare two different goal functions (2.32).

The $A^\star$ algorithms consist of the following steps [48]:

1. Inclusion of the initial architecture N_{arch_0} to the space of the network architectures $\mathcal{N}_{\mathrm{arch}}$.
2. Calculation of the goal function $f(N_{\mathrm{arch}_0})$.
3. Obtaining the network architecture $N^\star_{\mathrm{arch}}$ characterized by lower value of the goal function from the set $\mathcal{N}_{\mathrm{arch}}$.
4. If $N^\star_{\mathrm{arch}} \in \mathcal{G}$ then the algorithm is terminated and the architecture $N^\star_{\mathrm{arch}}$ constitute the solution.
5. Inclusion of the architecture $N^\star_{\mathrm{arch}}$ to the set $\mathcal{B}$.
6. Calculation of the goal functions for all successors of the network architecture $N^\star_{\mathrm{arch}}$, not included in the set $\mathcal{N}_{\mathrm{arch}}$ or $\mathcal{B}$.
7. Inclusion of all successors to the set $\mathcal{N}_{\mathrm{arch}}$ and return to step 2.

The $A^\star$ algorithm is an efficient tool in the optimisation of the neural networks architectures tasks. In the work [112] the supremacy of such algorithm over the cascade correlation algorithm has been presented. Moreover, it is shown that the successors in the subsequent iterations of $A^{\star}$ algorithm deliver the values of parameters for which the value of $J_{\mathcal{V}}$ is better then in the previous iterations:

$$\forall N'_{\text{arch}} \in \Xi(N_{\text{arch}}) \exists \boldsymbol{p}' : J_{\mathcal{V}}(\boldsymbol{y}, \boldsymbol{y}(N'_{\text{arch}}, \boldsymbol{p}')) \leq J_{\mathcal{V}}(\boldsymbol{y}, \boldsymbol{y}(N_{\text{arch}}, \boldsymbol{p}^{\star})). \quad (2.36)$$

Unfortunately, the disadvantage of $A^{\star}$ algorithm is high computational cost whereas the number of successors rapidly increases especially in the case of complex neural architectures.

2.4 GMDH in Design of Neural Networks

The concept of the GMDH approach relies on replacing the complex neural model by the set of hierarchically connected partial models. The model is obtained as a result of neural network structure synthesis with the application of the GMDH algorithm [69, 70, 71, 74, 116, 117]. The synthesis process consists of a partial model structure selection and parameters estimation. The parameters of each partial model (a neuron) are estimated separately. In the next step of the process synthesis, the partial models are evaluated, selected and included to newly created neuron layers (cf. Fig. 2.6). During the network synthesis new layers are added to the network. The process of network synthesis leads to the evolution of the resulting model structure to obtain the best quality approximation of real system output signals. The process is completed when the optimal degree of network complexity is achieved.

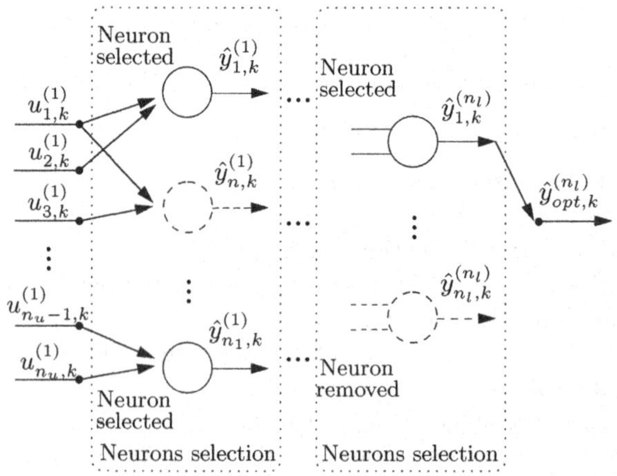

Fig. 2.6. Synthesis of the GMDH model

2.4.1 Synthesis of MISO GMDH Neural Network

Originally, the GMDH algorithm was developed for synthesis of the Multi-Input Single-Output (MISO) neural models [70]. In such case, based on the k-th measurement of the system inputs $\boldsymbol{u}_k \in \mathbb{R}^{n_u}$, the GMDH network grows its first layer of neurons. It is assumed that all possible couples of inputs from $u_{1,k}^{(l)}, ..., u_{n_u,k}^{(l)}$, belonging to the training data set $\mathcal{T}$, constitute the stimulation which results in the formation of the neurons outputs $\hat{y}_{n,k}^{(l)}$:

$$\hat{y}_{n,k}^{(l)} = f(\boldsymbol{u}_k) = f(u_{1,k}^{(l)}, \dots, u_{n_u,k}^{(l)}), \tag{2.37}$$

where l is the layer number of the GMDH network and n is the neuron number in the l-th layer.

The GMDH approach allows for much freedom in defining an neuron transfer function $f(\cdot)$ (e.g., tangent or logarithmic functions) [70, 118]. The original GMDH algorithm developed by Ivakhnenko [119] is based on linear or second-order polynomial transfer functions, such as:

$$f(u_{i,k}, u_{j,k}) = \hat{p}_0 + \hat{p}_1 u_{i,k} + \hat{p}_2 u_{j,k} + \hat{p}_3 u_{i,k} u_{j,k} + \hat{p}_4 u_{i,k}^2 + \hat{p}_5 u_{j,k}^2. \tag{2.38}$$

In this case, after network synthesis, the general relation between the model inputs and output $\hat{y}_k$ can be described in the following way:

$$\hat{y}_k = f(\boldsymbol{u}_k) = \hat{p}_0 + \sum_{i=1}^{n_u} \hat{p}_i u_{i,k} + \sum_{i=1}^{n_u} \sum_{j=1}^{n_u} \hat{p}_{ij} u_{i,k} u_{j,k} +, ...,. \tag{2.39}$$

From the practical point of view, (2.39) should be not too complex because it may complicate the learning process and extend computation time. In general, in the case of the identification of static non-linear systems, the partial model can be described as follows:

$$\hat{y}_{n,k}^{(l)} = f\left(\left(\boldsymbol{r}_{n,k}^{(l)}\right)^T \hat{\boldsymbol{p}}_n^{(l)}\right), \tag{2.40}$$

where $f(\cdot)$ denotes a non-linear invertible activation function, i.e., there exists $f^{-1}(\cdot)$. Moreover, $\boldsymbol{r}_{n,k}^{(l)} = \boldsymbol{d}\left([u_{i,k}^{(l)}, u_{j,k}^{(l)}]^T\right)$, $i, j = 1, \dots, n_u$ and $\hat{\boldsymbol{p}}_n^{(l)} \in \mathbb{R}^{n_p}$ are the regressor and parameter vectors, respectively, and $\boldsymbol{d}(\cdot)$ is an arbitrary bivariate vector function, e.g., $\boldsymbol{d}(\boldsymbol{x}) = [x_1^2, x_2^2, x_1 x_2, x_1, x_2, 1]^T$, that corresponds to the bivariate polynomial of the second degree.

The number of neurons in the first layer of the GMDH network depends on the number of the external inputs n_u:

$$\begin{cases} \hat{y}_{1,k}^{(1)} &= f(u_{1,k}^{(1)}, u_{2,k}^{(1)}, \hat{\boldsymbol{p}}_{1,2}) \\ \hat{y}_{2,k}^{(1)} &= f(u_{1,k}^{(1)}, u_{2,k}^{(1)}, \hat{\boldsymbol{p}}_{1,2}) \\ & \dots \\ \hat{y}_{n_y,k}^{(1)} &= f(u_{n_u-1,k}^{(1)}, u_{n_u,k}^{(1)}, \hat{\boldsymbol{p}}_{n_u-1,n_u}) \end{cases} \tag{2.41}$$

where $\hat{\boldsymbol{p}}_{1,2}, \hat{\boldsymbol{p}}_{1,3}, \ldots, \hat{\boldsymbol{p}}_{n_u-1,n_u}$ are the estimates of the network parameters and should be obtained during the identification process.

In order to estimate the unknown parameters $\boldsymbol{p}$ the techniques for the parameters estimation of linear-in-parameter models can be used, e.g., the LMS [69, 70]. It follows from the facts that the parameters of each partial models are estimated separately and the neuron's activation function $f(\cdot)$ fulfills the following conditions:

1. $f(\cdot)$ is continuous and bounded, i.e.,
$$\forall\, x \in \mathbb{R}\ : a < f(x) < b. \tag{2.42}$$
2. $f(\cdot)$ is monotonically increasing, i.e.,
$$\forall\, x, y \in \mathbb{R}\ : x \leq y \text{ iff } f(x) \leq f(y). \tag{2.43}$$
3. $f(\cdot)$ is invertible, i.e., there exists $f^{-1}(\cdot)$.

The advantage of this approach is a simple computation algorithm that gives good results even for small sets of measuring data. After the estimation, the parameters are "frozen" during the further network synthesis.

In the next stage of the GMDH network synthesis, the validation data set $\mathcal{V}$, not employed during the parameters estimation phase, is used to calculate a processing error of each partial model in the current l-th network layer. The processing error can be calculated with the application of the evaluation criteria such as: the Final Prediction Error (FPE), Akaike Information Criterion (AIC) or F-test [23, 118, 117]. Based on the defined evaluation criterion it is possible to select the best-fitted neurons in the layer. The selection methods in the GMDH neural networks play the role of a mechanism of structural optimisation at the stage of constructing a new layer of neurons. During the selection, neurons which have too large value of the evaluation criterion $Q(\hat{y}_{n,k}^{(l)})$, are rejected.

A few methods of performing the selection procedure can be applied [118]. One of the most often used is the constant population method. It is based on a selection of g neurons, which evaluation criterion $Q(\hat{y}_{n,k}^{(l)})$ reaches the smallest values. The constant g is chosen in an empirical way and the most important advantage of this method is its simplicity of implementation. Unfortunately, the constant population method has very restrictive structure evolution possibilities. One way out of this problem is the application of the optimal population method. This approach is based on rejecting the neurons which value of the evaluation criterion is larger than the arbitrarily determined threshold e_h. The threshold is usually selected for each layer in an empirical way depending on the considered task. The difficulty with the selection of the threshold results in the fact that the optimal population method is not applied too often. One of the most interesting ways of performing the selection procedure is the application of the method based on the soft selection approach. An outline of the Soft Selection Method (SSM) [71] is as follows:

Input : The set of all n_y neurons in the l-th layer, n_j – the number of opponent neurons, n_w – the number of winnings required for n-th neuron selection.
Output : The set of neurons after selection.

1. Calculate the evaluation criterion $Q(\hat{y}^{(l)}_{n,k})$ for $n = 1, \ldots, n_y$ neurons;
2. Conduct a series of n_y competitions between each n-th neuron in the layer and n_j randomly selected neurons (the so-called opponent) from the same layer. The n-th neuron is the so-called winner neuron when:

$$Q(\hat{y}^{(l)}_{n,k}) \leq Q(\hat{y}^{(l)}_{j,k}), \quad j = 1, \ldots, n_j,$$

   where $\hat{y}^{(l)}_{j,k}$ denotes a signal generated by the opponent neuron;
3. Select the neurons for the $(l + 1)$-th layer with the number of winnings bigger than n_w (the remaining neurons are removed).

The property of soft selection follows from the specific series of competitions. It may happen that the potentially unfitted neuron is selected. Everything depends on its score in the series of competition. The main advantage of such approach in comparison to other selection methods is that it is possible to use potentially unfitted neurons which in the next layers may improve the quality of the model. Moreover, if the neural network is not fitted perfectly to the identification data set, it is possible to achieve a network which possess better generalization abilities. One of the most important parameters, which should be chosen in the selection process, is the number of n_j opponents. A bigger value of n_j makes the probability of the selection of a neuron with a small quality index lower. In this way, in an extreme situation, when $n_j \gg n_y$, the SSM will behave as the constant population method, which is based on the selection only of the best fitted neurons. Some experimental results performed on a number of selected examples indicate that the SSM makes it possible to obtain a more flexible network structure. Another advantage, in comparison to the optimal population method, is that an arbitrary selection of the threshold is avoided. Instead, we have to select a number of winnings n_w. It is, of course, a less sophisticated task.

After the selection procedure, the outputs of the selected neurons become the inputs to other neurons in the next layer. Similarly, new neurons in the next layers of the network are created. During the synthesis of the GMDH network, the number of layers suitably increases. Each time a new layer is added, new neurons are introduced. The synthesis of the GMDH network is completed when so-called the optimum criterion is achieved. The idea of this criterion relies on the determination of the quality index $Q(\hat{y}^{(l)}_{n,k})$ for all N neurons included in the l layer. $Q^{(l)}_{min}$ represents the processing error for the best neuron in this layer:

$$Q_{min}^{(l)} = \min_{n=1,\ldots,N} Q(\hat{y}_{n,k}^{(l)}). \tag{2.44}$$

The values $Q(\hat{y}_{n,k}^{(l)})$ can be determined with the application of the defined evaluation criterion, which was used in the selection process. The values $Q_{min}^{(l)}$ are calculated for each layer in the network. The optimum criterion is achieved when the following condition occurs:

$$Q_{opt}^{(L)} = \min_{l=1,\ldots,L} Q_{min}^{(l)}. \tag{2.45}$$

$Q_{opt}^{(L)}$ represents the processing error for the best neuron in the network, which generates the model output. In other words, when additional layers do not improve the performance of the network, the synthesis process is stopped.

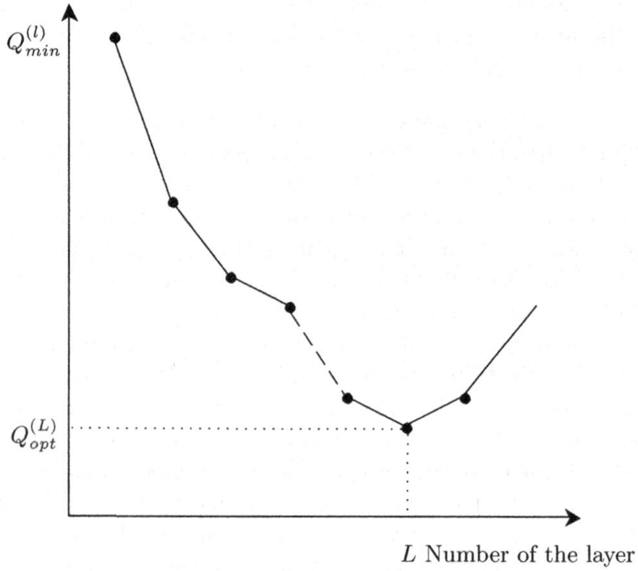

Fig. 2.7. Termination of the MISO GMDH model synthesis

To obtain the final structure of the network, all unnecessary neurons are removed, leaving only those which are relevant to the computation of the model output. The procedure of removing unnecessary neurons is the last stage of the synthesis of the GMDH neural network.

2.4.2 Synthesis of MIMO GMDH Neural Network

The assumptions of the GMDH approach presented in Sect. 2.4.1 lead to the formation of the neural network of a multi input $u_{1,k}, u_{2,k}, \ldots, u_{n_u,k}$ and a single output y_k. However, systems of the multi input $u_{1,k}, u_{2,k}, \ldots, u_{n_u,k}$ and the multi output $y_{1,k}, \ldots, y_{r,k}, \ldots, y_{n_y,k}$ are found in practical applications most often. The synthesis of the MIMO model [120] can be performed in a similar way as in the case of MISO models. In the first step of the network synthesis, based on all combinations of the inputs, the estimates of $\hat{y}_{1,k}^{(1)}$ system output $y_{1,k}^{(1)}$ is obtained. Next, based on the same combinations of the input signals, the remaining estimates $\hat{y}_{2,k}^{(1)}, \ldots, \hat{y}_{n_y,k}^{(1)}$ outputs $y_{2,k}^{(1)}, \ldots, y_{n_y,k}^{(1)}$ are obtained.

$$\left\{\begin{array}{ll} \hat{y}_{1,1,k}^{(1)} & \ldots f(u_{1,k}^{(1)}, u_{2,k}^{(1)}) \\ \hat{y}_{1,2,k}^{(1)} & \ldots f(u_{1,k}^{(1)}, u_{3,k}^{(1)}) \\ & \ldots \\ \hat{y}_{1,n_N,k}^{(1)} & \ldots f(u_{n_u-1,k}^{(1)}, u_{n_u,k}^{(1)}) \\ & \vdots \\ \hat{y}_{n_y,1,k}^{(1)} & \ldots f(u_{1,k}^{(1)}, u_{2,k}^{(1)}) \\ \hat{y}_{n_y,2,k}^{(1)} & \ldots f(u_{1,k}^{(1)}, u_{3,k}^{(1)}) \\ & \ldots \\ \hat{y}_{n_y,n_N,k}^{(1)} & \ldots f(u_{n_u-1,k}^{(1)}, u_{n_u,k}^{(1)}) \end{array}\right. \tag{2.46}$$

The first layer of the GMDH network has the structure shown in Fig. 2.8.

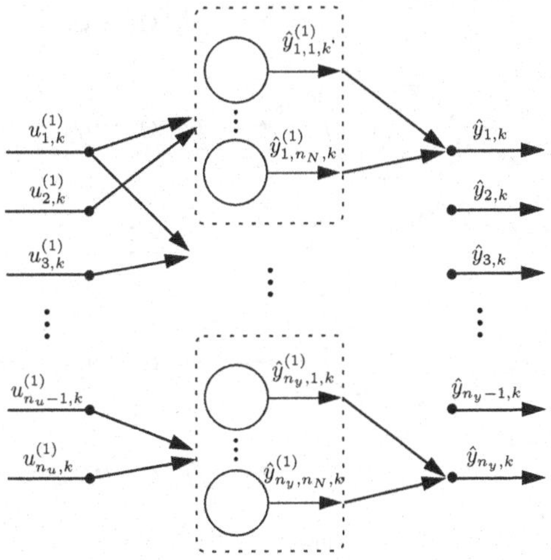

Fig. 2.8. First layer of the MIMO GMDH model

The definition of the evaluation criterion $Q_r(\hat{y}_{r,n}^{(l)})$ of neurons is a preliminary task in designing the GMDH network [118]. It allows any neuron to define the quantity of a processing error of each output. The independent evaluation of any processing errors $Q_1, \ldots, Q_r, \ldots, Q_{n_y}$ is performed after the generation of each layer of neurons. Moreover, based on the defined evaluation criterion it is possible to make the selection of neurons in the layer:

$$\begin{cases} Q_1(\hat{y}_{1,1}^{(l)}) & \dots Q_{n_y}(\hat{y}_{1,1}^{(l)}) \\ Q_1(\hat{y}_{1,2}^{(l)}) & \dots Q_{n_y}(\hat{y}_{1,2}^{(l)}) \\ & \dots \\ Q_1(\hat{y}_{1,n_N}^{(l)}) & \dots Q_{n_y}(\hat{y}_{1,n_N}^{(l)}) \\ & \dots \\ Q_1(\hat{y}_{n_y,1}^{(l)}) & \dots Q_{n_y}(\hat{y}_{n_y,1}^{(l)}) \\ Q_1(\hat{y}_{n_y,2}^{(l)}) & \dots Q_{n_y}(\hat{y}_{n_y,2}^{(l)}) \\ & \dots \\ Q_1(\hat{y}_{n_y,n_N}^{(l)}) & \dots Q_{n_y}(\hat{y}_{n_y,n_N}^{(l)}) \end{cases} \tag{2.47}$$

The selection of the best performing neurons in the terms of their processing accuracy in the layer is realized with the application of selection methods [71], before the formed layer is added to the network. According to the chosen selection method, elements that introduce too big processing error of each output $y_1, \dots, y_{n_y}$ are removed (cf. Fig. 2.9).

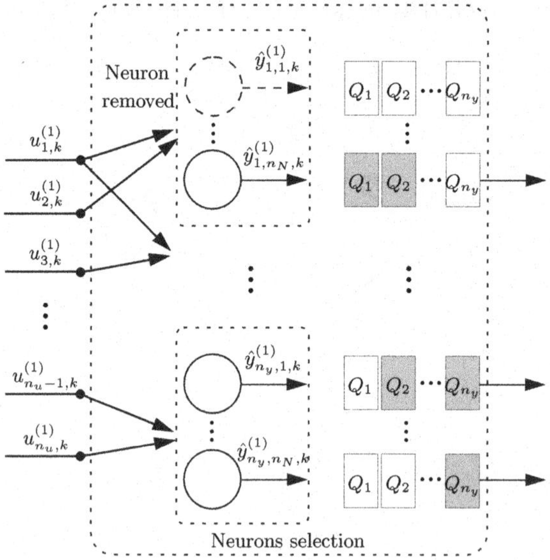

Fig. 2.9. Selection in the MIMO GMDH model

The effectiveness of a neuron in processing at least one output signal is sufficient to leave the neuron in the network. Based on all selected neurons, a new layer is created. The parameters of the neurons in a newly created layer are „frozen" during further network synthesis. The outputs of the selected neurons become the inputs to other neurons in the next layer. During the synthesis of the next layers, all selected outputs from the previous layer must be used to generate each output $y_1, \ldots, y_{n_y}$. It follows from the fact that in real industrial systems outputs are usually correlated, so the output $\hat{y}_{r,n}^{(l)}$ should be obtained based on all potential outputs $\hat{y}_{1,1}^{(l-1)}, \ldots, \hat{y}_{n_y,n_N}^{(l-1)}$. The synthesis of the GMDH network is completed when the optimum criterion is obtained (cf. Fig. 2.10).

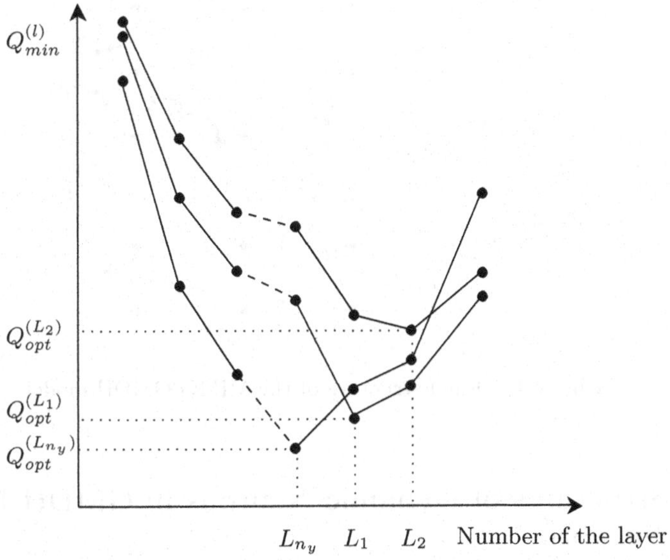

Fig. 2.10. Termination of the MIMO GMDH model synthesis

The idea of this criterion is the determination of the quality index $Q_r(\hat{y}_{r,n}^{(l)})$, for all n_N neurons included in the l layer, for each system output $y_1, \ldots, y_{n_y}$ independently:

$$Q_{r,min}^{(l)} = \min_{n=1,\ldots,n_N} Q_r(\hat{y}_{r,n}^{(l)}) \text{ for } r = 1, \ldots, n_y. \tag{2.48}$$

The $Q_{r,min}^{(l)}$ represents the processing error of the best neuron in this layer for the output y_r. The values $Q_r(\hat{y}_{r,n}^{(l)})$ can be determined with the application of the defined evaluation criterion used in the selection process. The values $Q_{r,min}^{(l)}$ are calculated for each layer in the network. The optimum criterion for the MIMO GMDH network is obtained when the following condition occurs:

$$Q_{opt}^{(L_r)} = \min_{l=1,\ldots,L_r} Q_{r,min}^{(l)} \quad \text{for} \quad r = 1, \ldots, n_y. \tag{2.49}$$

In other words, the synthesis of the network is completed when each of the calculated values $Q^{(l)}_{r,min}$ reaches the minimum. The output y_r is connected to the output of the neuron for which $Q^{(l)}_{r,min}$ achieves the least value. The particular minimum could occur in different stages of the network synthesis. This is why in the multi-output network, the outputs of the resulting structure are usually in different layers (cf. Fig. 2.11).

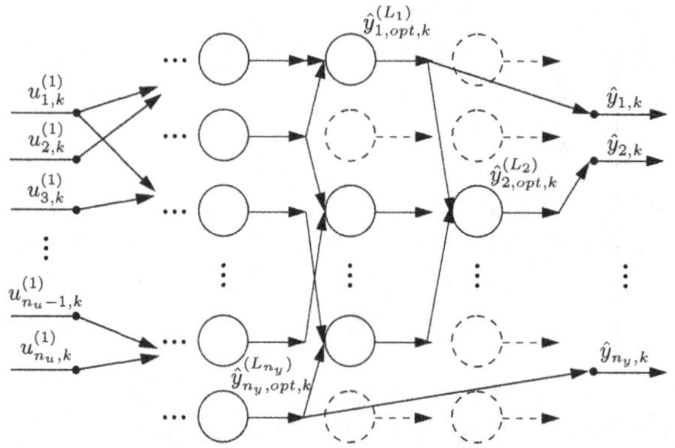

Fig. 2.11. Final structure of the MIMO GMDH model

2.5 Structures of Dynamic Neurons in GMDH Neural Networks

2.5.1 Dynamic Neuron with IIR Filter

The GMDH approach allows much freedom in defining the partial model structure. In general, in the case of the identification of the static non-linear systems, the partial model can be described as (2.40). As it was already mentioned in Sect. 2.4.1 the neuron (2.40) allows to use the techniques for the parameters estimation of linear-in-parameter models. Indeed, since $f(\cdot)$ is invertible, the neuron described by (2.40) can be relatively easily transformed into a linear-in-parameter one. Such neuron can be used for the identification of the non-linear static systems. Unfortunately, the most of the industrial systems are dynamic in its nature [1, 11]. The application of the static neural network will result in a large model uncertainty. Thus, during the system identification it seems desirable to employ the models, which can represent the dynamic of the system. In the case of the classical neural network, for example in the MLP, the problem of modelling of the dynamics is solved by the introduction of additional input signals. The details of this approach are presented in Sect. 2.2. Unfortunately, the described approach can not be applied into the GMDH neural network easily, because the GMDH network is constructed through the gradual connection of the partial models. The introduction of the global output feedback lines complicates the synthesis of the network. On the other hand, the behaviour of each partial model should reflect the behaviour of the identified system. It follows from the rule of the GMDH algorithm that the parameters of each partial model are estimated in such a way that their output signals are the best approximation of the real system output. In this situation, the partial model should have an ability to represent the dynamics. One way out of this problem is to use dynamic neurons. The neural networks constructed by the connection of such kind of neurons are called locally recurrent networks. Due to the introduction of different local feedbacks to the classical McCulloch-Pitts neuron (2.40), it is possible to achieve a few types of dynamic neurons, e.g., with local activation feedback [121], with local synapse feedback [122], with output feedback [123] and neuron with memory [124].

The neuron with the local activation feedback [121] is described by the following equations:

$$\varphi_k = \sum_{i=1}^{n_p} p_i r_{i,k} + \sum_{j=1}^{n_b} b_j \varphi_{k-j}, \tag{2.50}$$

and

$$\hat{y}_k = f(\varphi_k), \tag{2.51}$$

where $r_{i,k}$, $i = 1, 2, \ldots, n_p$ denotes the inputs of neuron, p_i are input parameters, φ_k is the activation potential, b_j, $j = 1, 2, \ldots, n_b$ are the coefficients which determine the feedback intensity of φ_{k-j} and $f(\cdot)$ is a non-linear activation function. Note that the right side of sum in (2.50) can be interpreted as the Finite Impulse Response (FIR) filter.

The neuron with the local synapse feedback [122] has a structure in which the synapse with a linear transfer function with poles and zeros is applied – an Infinite Impulse Response (IIR) filter:

$$\hat{y}_k = f\left(\sum_{i=1}^{n_p} G_i(z^{-1}) r_{i,k}\right), \tag{2.52}$$

where:

$$G_i(z^{-1}) = \frac{\sum_{j=0}^{n_b} b_j z^{-1}}{\sum_{j=0}^{n_a} a_j z^{-1}}. \tag{2.53}$$

where: $G_i(z^{-1})$ is the transfer function, where b_j, $j = 0, 1, \ldots, n_b$, and a_j, $j = 0, 1, \ldots, n_a$ are its zeros and poles which should be estimated. Note that the inputs of the neuron may be taken from the outputs of the previous layer, or from the output of the neuron. If they are derived from the previous layer, then it is a local synapse feedback. On the other hand, if they are derived from the output y_k it is a local output feedback. Moreover, the local activation feedback is a special case of the local synapse feedback architecture. In this case, all synaptic transfer functions have the same denominator and only one zero.

The neuron with the output feedback [123] opposite to the local synapse and local activation feedback has the feedback after the non-linear activation submodule:

$$\hat{y}_k = f\left(\sum_{i=1}^{n_p} p_i r_{i,k} + \sum_{j=1}^{n_c} c_j \hat{y}_{k-j}\right), \tag{2.54}$$

where c_j for $j = 1, 2, \ldots, n_c$ represents the coefficients which determine the feedback intensity of the neuron output $\hat{y}_{k-j}$. In this architecture, the output of the neuron is filtered by the FIR filter, and it is added to the set of neuron inputs.

The dynamic neuron with the memory [124] can accumulate information regarding past activations of neurons. The behaviour of such a neuron is characterized by the relation:

$$y_k = f\left(\sum_{i=1}^{n_p} p_i r_{p,k} + \sum_{i=1}^{n_p} s_i z_{i,k}\right), \tag{2.55}$$

where:

$$z_{i,k} = \alpha_p r_{p,k-1} + (1 - \alpha_p) z_{p,k-1}, \tag{2.56}$$

where α denotes constant coefficient, z_p for $p = 1, 2, \ldots, n_p$ are the outputs of the memory neuron from the previous layer, and s_p for $p = 1, 2, \ldots, n_p$ are estimated parameters. The neuron with the memory can be perceived as a special case of the generalized local output feedback architecture. It has a feedback transfer function only with one pole.

The feed-forward structure of networks consisting of the dynamic neurons seems to make the training process easier in comparison to the global recurrent network presented in Sect. 2.2. On the other hand, the introduction of the dynamic neurons increases the parameter space significantly. This drawback together with the non-linear and multi-modal properties of the dynamic neuron implies that the parameters estimation still becomes complex.

In order to overcome these drawbacks it is possible to use another kind of a dynamic neuron model [74]. Dynamics in this neuron is realized by the introduction of a linear dynamic system - the IIR filter. In this way, each neuron in the network reproduces the output signal based on the past values of its inputs and outputs. Such a neuron model (Fig. 2.12) consists of two submodules: the filter module and the activation module.

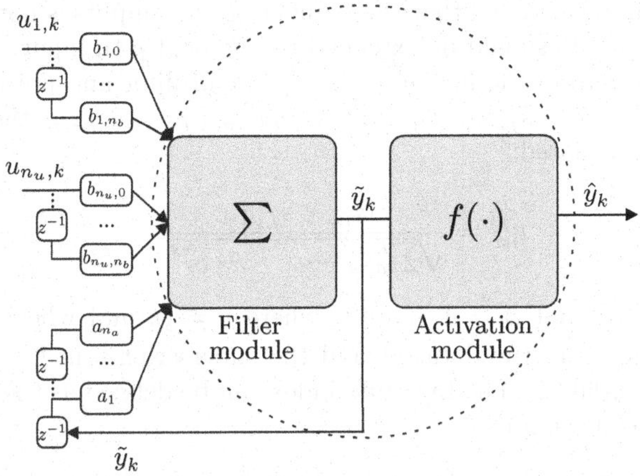

Fig. 2.12. Dynamic neuron model with the IIR filter

The behaviour of the filter module is described by the following equation:

$$\tilde{y}_k = -a_1\tilde{y}_{k-1} - \ldots - a_{n_a}\tilde{y}_{k-n_a} + \boldsymbol{b}_0^T\boldsymbol{u}_k + \boldsymbol{b}_1^T\boldsymbol{u}_{k-1}+,\ldots,+\boldsymbol{b}_{n_b}^T\boldsymbol{u}_{k-n_b}, \quad (2.57)$$

or equivalently,

$$\tilde{y}_k = (\boldsymbol{r}_k)^T\hat{\boldsymbol{p}}. \quad (2.58)$$

where $\boldsymbol{r}_k = [-\tilde{y}_{k-1},\ldots,-\tilde{y}_{k-n_a},\boldsymbol{u}_k,\boldsymbol{u}_{k-1},\ldots,\boldsymbol{u}_{k-n_b}]$ represents the regressor and $\hat{\boldsymbol{p}} = [a_1,\ldots,a_{n_a},\boldsymbol{b}_0,\boldsymbol{b}_1,\ldots,\boldsymbol{b}_{n_b}]$ denotes the filter parameters. The filter output is used as the input for the activation module:

$$\hat{y}_k = f(\tilde{y}_k). \quad (2.59)$$

The main advantage of the proposed dynamic neuron with the IIR filter is the possibility of an application of the techniques for the parameters estimation of linear-in-parameter models when the activation function $f(\cdot)$ is invertible [71].

2.5.2 Order Selection of Dynamic Neuron

The creation of high quality of the dynamic GMDH neural model requires the appropriate order of the dynamic neuron selection. This problem can be solved by the application of the Lipschitz index approach based on the so-called the Lipschitz quotients [25]. The concept of the proposed method is based on the following strategy. Let us assume that the value $\widetilde{s}_{j,k}$ in (2.66) depends on p inputs. If two measurements $\imath$ and $\wp$ are close to each other in the space inputs φ_{p-1}, then it can be expected that related to them outputs $\widetilde{s}_{j,\imath}$ and $\widetilde{s}_{j,\wp}$ should have close values. If the same two measurements $\imath$ and $\wp$ in the space of inputs φ_p are far away, and if one or few of inputs of p are missing, then the values of $\widetilde{s}_{j,\imath}$ and $\widetilde{s}_{j,\wp}$ should have different values. In this situation, it can be assumed that the number of φ_{p-1} inputs is not sufficient, and the set of inputs should be extended by the next p-th input.

Thus, starting from the first input $\varphi_1 = r_{i,k}$ of filter and finishing on all inputs $\varphi_1 = r_{i,k}$, $\varphi_2 = \widetilde{s}_{j,k-1}$, ..., $\varphi_{q-1} = r_{i,k-n_b}$, $\varphi_q = \widetilde{s}_{j,k-n_a}$ the Lipschitz quotients are calculated:

$$l_{\imath\wp}^{(p)} = \frac{|\widetilde{s}_{j,\imath} - \widetilde{s}_{j,\wp}|}{\sqrt{\sum_{n=1}^{p}(\varphi_{n,\imath} - \varphi_{n,\wp})^2}}, \tag{2.60}$$

for $\imath = 1, \ldots, n_{\mathcal{T}}$ and $\wp = 1, \ldots, n_{\mathcal{T}}$, where $\imath \neq \wp$, and where $n_{\mathcal{V}}$ is the number of samples in the data set and the superscript p in $l_{\imath\wp}^{(p)}$ stands for the number of inputs φ. The Lipschitz index can be defined as the maximum occurring Lipschitz quotient:

$$l^{(p)} = \max_{\imath,\wp,(\imath\neq\wp)} (l_{\imath\wp}^{(p)}). \tag{2.61}$$

In a general case, the index is large if several inputs $\varphi_1, \varphi_2, \ldots, \varphi_p$ are missing and is small otherwise. In other words, the correct order of the dynamic neuron can be detected when the index Lipschitz stops to decrease (Fig. 2.13) (for $p = 6$).

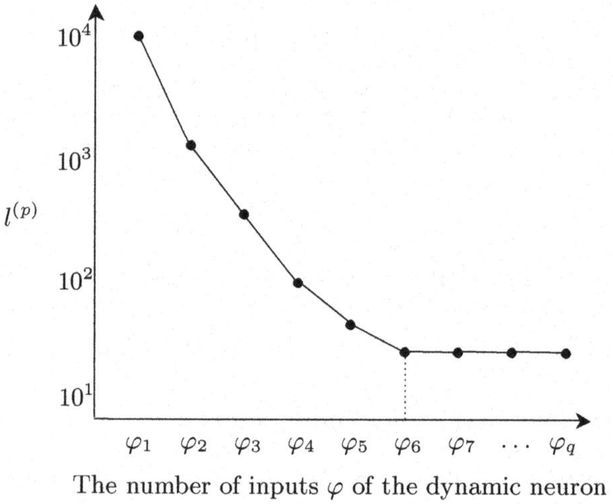

Fig. 2.13. Influence of number of neuron inputs on Lipschitz quotients

The presented approach has two main advantages. It allows to choose the dynamic neuron structure only on the basis of the measurement data set before the parameters estimation. Moreover, it has no restrictions as far as the structure of the neuron is concerned. It can be applied for any type of dynamic neurons.

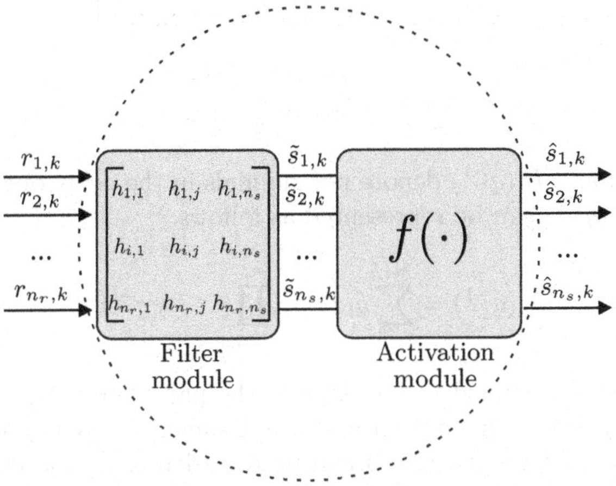

Fig. 2.14. Dynamic neuron model in the polar coordinates

2.5.3 Dynamic Neuron in Polar Coordinates

The main advantage of locally recurrent networks constructed with the application of the dynamic neurons is that their stability can be proved easier opposite to the global recurrent networks. As a matter of the fact, the stability of the network only depends on the stability of neurons. In this section the dynamic neuron in the polar representation is defined. The neuron parameters have a physical interpretation related to the position and bandwidths of spectral peaks. The neuron stability can be easily attained due to the clear relation between the stability and the poles, i.e., they should lie within a unit circle. The proposed dynamic neuron [76] consists of two submodules: the filter module and activation module (cf. Fig. 2.14). The behavior of the filter module is described by the following equation:

$$h_{i,j}(q^{-1}) = \frac{\tilde{s}_{j,k}(q^{-1})}{r_{i,k}(q^{-1})} = \frac{B_i(q^{-1})}{A_j(q^{-1})}, \tag{2.62}$$

where $i = 1, \ldots, n_r$, $j = 1, \ldots, n_s$ and:

$$A_j(q^{-1}) = 1 + \ldots + a_{j,l}q^{-l} + \ldots + a_{n_y,n_a}q^{-n_a}, \tag{2.63}$$

$$B_i(q^{-1}) = b_{i,0} + \ldots + b_{i,m}q^{-m} + \ldots + b_{n_u n_b}q^{-n_b}, \tag{2.64}$$

where: $l = 1, \ldots, n_a$, $m = 1, \ldots, n_b$ and $\hat{\boldsymbol{p}} = [a_{1,1}, \ldots, a_{n_y,n_a}, b_{10}, \ldots, b_{n_u,n_b}]^T = [\boldsymbol{a}, \boldsymbol{b}]^T$, represents the vector of the parameters estimate. The filter output $\tilde{s}_{j,k}$ is used as the input for the activation module:

$$\hat{s}_{j,k} = f\left(\tilde{s}_{j,k}\right), \tag{2.65}$$

where $f(\cdot)$ denotes a non-linear activation function, while:

$$\tilde{s}_{j,k} = \frac{B_i(q^{-1})}{A_j(q^{-1})} r_{i,k}, \tag{2.66}$$

where $A_j(q^{-1})$ and $B_i(q^{-1})$ denote polynomials in the delay operator q. The polynomial $A_j(q^{-1})$ can be represented as follows:

$$A_j(q^{-1}) = \sum_{l=0}^{n_a} a_l q^{-l} = \prod_{l=1}^{n_a} (1 - \lambda_l q^{-l}). \tag{2.67}$$

Coefficients a_l are assumed to be real and the parameters λ_l, $l = 1, \ldots, n_a$, represent the poles of the system and are assumed to have magnitude less than unity. The polynomial (2.67) can be regarded as a cascade of the first and second order filter sections when n_a is an odd number:

$$A_j(q^{-1}) = F_l(q^{-1}) \prod_{l=1}^{(n_a-1)/2} S_l(q^{-1}), \tag{2.68}$$

and:

$$A_j(q^{-1}) = \prod_{l=1}^{(n_a-1)/2} S_l(q^{-1}), \tag{2.69}$$

otherwise, where:

$$F_l(q^{-1}) = (1 - \zeta_l q^{-1}), \tag{2.70}$$

and:

$$S_l(q^{-1}) = (1 - \lambda_l q^{-1})(1 - \lambda_l^* q^{-1}), \tag{2.71}$$

where λ_l and λ_l^*, for $l = 1, \ldots, n_a$ are the l-th pair of complex conjugate roots i.e $\lambda = \alpha + \beta i$ and $\lambda^* = \alpha - \beta i$, and ζ_l, is the l-th real root.

The problem is to obtain the estimates of poles λ_l and λ_l^* directly from the identification data. Each pair of complex conjugate roots of the polynomial $A_j(q^{-1})$ can be defined in the following form:

$$\lambda_l = \rho_l e^{i\omega_l}, \quad i = \sqrt{-1}, \tag{2.72}$$

where ρ_l and ω_l are the radius and positive angle of the l-th pair of the complex conjugate roots. Now, the unknown parameters of $A_j(q^{-l})$ are defined as follows:

$$\boldsymbol{a} = [\boldsymbol{\zeta}_l^T, \boldsymbol{\rho}_l^T, \boldsymbol{\omega}_l^T]^T, \tag{2.73}$$

where:

$$\boldsymbol{\zeta}_l = [\zeta_1, \zeta_2, \ldots, \zeta_{n_a}]^T, \tag{2.74}$$

$$\boldsymbol{\rho}_l = [\rho_1, \rho_2, \ldots, \rho_{n_a}]^T, \tag{2.75}$$

$$\boldsymbol{\omega}_l = [\omega_1, \omega_2, \ldots, \omega_{n_a}]^T. \tag{2.76}$$

The second-order filter section (2.71) can be written in the polar coordinates:

$$\boldsymbol{S}_l(q^{-2}) = 1 - 2\boldsymbol{\rho}_l \cos(\boldsymbol{\omega}_l)q^{-1} + \boldsymbol{\rho}_l^2 q^{-2}. \tag{2.77}$$

where the roots have the following form:

$$\boldsymbol{\lambda}_l = (\cos(\boldsymbol{\omega}_l) + |\sin(\boldsymbol{\omega}_l)|i)\boldsymbol{\rho}_l, \tag{2.78}$$

$$\boldsymbol{\lambda}_l{}^* = (\cos(\boldsymbol{\omega}_l) - |\sin(\boldsymbol{\omega}_l)|i)\boldsymbol{\rho}_l. \tag{2.79}$$

The absolute value of the roots $(\lambda_l, \lambda_l^*)_{l=0}^{n_a}$ is given in the form:

$$|\boldsymbol{\lambda}_l| = |\boldsymbol{\lambda}_l^*| = \sqrt{(\cos(\boldsymbol{\omega}_l)^2 + |\sin(\boldsymbol{\omega}_l)|^2)\boldsymbol{\rho}_l^2} = \boldsymbol{\rho}_l. \tag{2.80}$$

The neuron model is asymptotically stable when all roots lie within the unit circle:

$$0 \leq \boldsymbol{\rho}_l \leq 1 - \delta, \text{and} |\boldsymbol{\zeta}_l| \leq 1 - \delta, \quad l = 1, \ldots, n_a, \tag{2.81}$$

where δ is a sufficiently small positive constant. As the functions sin and cos are periodical, the following additional constraint can be introduced:

$$0 \leq \boldsymbol{\omega}_l \leq 2\pi, \quad l = 1, \ldots, n_a. \tag{2.82}$$

Finally, the training of the dynamic neuron $\hat{\boldsymbol{p}} = [\boldsymbol{a}, \boldsymbol{b}]^T$ relies on the estimation of the vectors parameters $\hat{\boldsymbol{p}} = [\boldsymbol{\zeta}_l^T \boldsymbol{\rho}_l^T, \boldsymbol{\omega}_l^T, \boldsymbol{\zeta}_m^T \boldsymbol{\rho}_m^T, \boldsymbol{\omega}_m^T]^T$ with the constraints (2.81-2.82).

2.5.4 State-Space Dynamic Neuron

The weakness of the dynamic in the polar coordinate representation described in Sect.2.5.3 is relatively high computational cost during its parameters estimation. It results from the large number of parameters especially in the case of high order of dynamic. This problem can be oppressive during the synthesis of the GMDH model consisting of several neurons. Moreover, all above mentioned dynamic neurons do not have a state-space description. In fact, the approaches trying to solve such a challenging problem can be rarely found in the literature [125, 126]. Unfortunately, these approaches do not allow to calculate the uncertainty of these models, what is necessary to apply in the robust fault detection schemes. In order to solve all these problems the dynamic neuron in the state-space representation is proposed in this section. The proposed dynamic neuron consists of the linear state-space module and activation module [61, 77] (cf. Fig. 2.15). The behavior of the linear state-space part of the dynamic neuron is described by the following equation:

$$\boldsymbol{z}_{k+1} = \boldsymbol{A}\boldsymbol{z}_k + \boldsymbol{B}\boldsymbol{r}_{i,k}, \tag{2.83}$$

$$\tilde{\boldsymbol{s}}_{i,j,k} = \boldsymbol{C}\boldsymbol{z}_k, \tag{2.84}$$

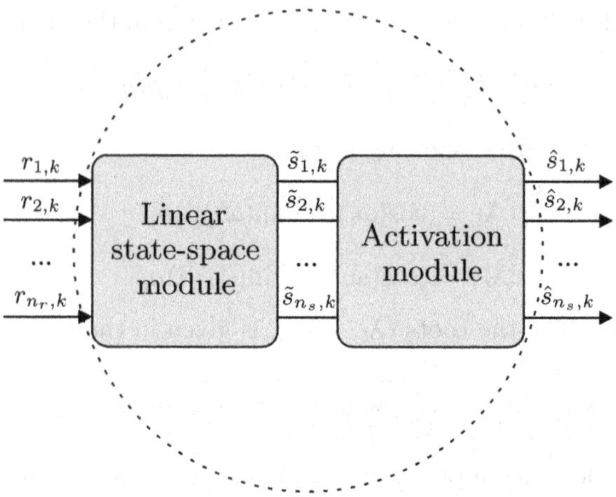

Fig. 2.15. Dynamic neuron model in the state-space representation

where $\boldsymbol{r}_{i,k} \in \mathbb{R}^{n_r}$ and $\tilde{\boldsymbol{s}}_{i,j,k} \in \mathbb{R}^{n_s}$ are the inputs and outputs of the linear state-space submodule of the dynamic neuron. $\boldsymbol{A} \in \mathbb{R}^{n_z \times n_z}$, $\boldsymbol{B} \in \mathbb{R}^{n_z \times n_r}$, $\boldsymbol{C} \in \mathbb{R}^{n_s \times n_z}$, $\boldsymbol{z}_k \in \mathbb{R}^{n_z}$, where n_z represents the order of the dynamics. Additionally, the matrix $\boldsymbol{A}$ has an upper-triangular form:

$$\boldsymbol{A} = \begin{bmatrix} a_{11} & a_{12} & \cdots & a_{1,n_z} \\ 0 & a_{22} & \cdots & a_{2,n_z} \\ \vdots & & \ddots & \vdots \\ 0 & 0 & \cdots & a_{n_z,n_z} \end{bmatrix}. \tag{2.85}$$

This means that the dynamic neuron is asymptotically stable iff:

$$\mid a_{i,i} \mid < 1, \quad i = 1, ..., n_z. \tag{2.86}$$

Moreover:

$$\boldsymbol{C} = \text{diag}(c_1, ..., c_{n_s}, \underbrace{0, ..., 0}_{n_z - n_s}). \tag{2.87}$$

The linear state-space submodule output is used as the input for the activation module:

$$\hat{\boldsymbol{s}}_{i,j,k} = \mathcal{F}(\tilde{\boldsymbol{s}}_{i,j,k}). \tag{2.88}$$

with $\mathcal{F}(\cdot) = [f_1(\cdot), ..., f_{n_s}(\cdot)]^T$ where $f_i(\cdot)$ denotes a non-linear activation function (e.g., a hyperbolic tangent).

The application of the dynamic neurons presented in Sect. 2.5.4 during the process of the GMDH network synthesis allows to obtain the dynamic GMDH neural model. This model has a cascade structure what follows from the fact that the neurons outputs constitute the neurons inputs in the subsequent layers. The neural model, which is the result of the cascade connection of the dynamic neurons, is asymptotically stable when each of neurons is asymptotically stable [127]. So, the fulfilment of the conditions (2.81-2.82) in the case of the dynamic neurons in the polar representation and condition (2.86) in the case of dynamic neurons in the state-space representation, during the process of estimation of particular dynamic neurons allows to obtain an asymptotically stable dynamic GMDH neural model.

2.6 Sources of GMDH Neural Models Uncertainty

The objective of the system identification is to obtain a mathematical description of a real system based on input-output measurements. The model obtained in this process should be high quality, otherwise it can not be applied in the FDI systems. The application of the GMDH approach to the model structure selection can improve the quality of the model but it can not eliminate the model uncertainty at all. Indeed, irrespective of the identification method used, there is always the problem of model uncertainty, i.e., the model-reality mismatch [10]. It is necessary to analyze each stage of the neural model synthesis process in order to identify each source of uncertainty. This knowledge makes it possible to reduce the model uncertainty or eventually calculate in the form which can be applied in the robust FDI schemes.

2.6.1 Neural Network Structure Errors

The errors in the structure of the network are important sources of uncertainty of the GMDH neural model. There are few reasons of the appearance of structural errors in the GMDH network, e.g., an unappropriate structure of elementary models, errors in the process of selection and caused by the evaluation criteria.

In the GMDH algorithm it is assumed that the single neuron is an elementary model which should approximate the identified system behaviour. In fact, this condition is difficult to satisfy. Indeed, in the first layer of the GMDH network the neurons are usually created only on the basis of two input variables selected from all inputs $\boldsymbol{u}_k$. In this case, the appropriate modeling of the system behaviour is quite difficult or even possible. Admittedly, in the subsequent layer of the GMDH network the neuron can connect the outputs of two neurons from the previous layer which are created on the basis of different inputs. In this way the neuron, which generates the output signal of the whole GMDH network, can connect all input signals. Unfortunately, it is not possible to exclude the case when during the process of selection only one neuron, which is connected to some input signal, is removed from the network. In this way, after the termination of the network synthesis the model not containing all inputs signal, can be obtained.

Moreover, the choice of the unappropriate structure of the elementary structure model can be a source of model uncertainty. It should be underlined that the elementary model should reflect the dynamic nature of the identified system. In order to solve this problem in Sect. 2.5 a few structures of the dynamic neurons are proposed. Moreover, not application of the approach of selection of the order of dynamic neuron can also increase the GMDH neural model uncertainty.

The next source of the structure errors can be caused by the selection methods applied during network synthesis. From the theoretical point of view, the selection method should ensure the choice of optimal structure of the network. Unfortunately, an unappropriate assumption of parameters influencing on the selection method can lead to the rejection of the neurons which should be included in the network. In order to prevent this situation the proposed in Sect. 2.4.1 the SSM should be applied during the GMDH network synthesis.

Furthermore, widely applied evaluation criteria of the partial models can be sources of the GMDH model uncertainty. Mueller and Lemke in [118] present a comprehensive table of the most common evaluation criteria used in the GMDH algorithm. Most often applied are the AIC and FPE, which are based on the statistics taking into consideration the complexity of partial models. The optimal structure of the partial model is obtained when the statistics has the minimal value. In the case of the AIC criterion statistics has the following general form:

$$Q_{\mathrm{AIC}}(\hat{y}_{n,k}^{(l)}) = n_{\mathcal{V}} \log J_{n_{\mathcal{V}}}(N_{\mathrm{arch}}) + \gamma(n_{\mathcal{V}}, n_p), \tag{2.89}$$

where $J_{n_{\mathcal{V}}}(N_{\mathrm{arch}})$ represents the goal function for the model architecture N_{arch}, e.g., $J_{n_{\mathcal{V}}}(N_{\mathrm{arch}}) = \frac{1}{n_{\mathcal{V}}} \sum_{k=1}^{n_{\mathcal{V}}} (y_k - \hat{\boldsymbol{y}}_{n,k}^{(l)})^2$, and $\gamma(n_{\mathcal{V}}, n_p)$ is the function of the number of the data samples $n_{\mathcal{V}}$ and number of partial model parameters n_p. The appropriate selection of (2.89) ensures its increase along with increase of the number of parameters and converge to zero along with increase of the data samples set. The selection of the function characterized by the above mentioned properties ensures an elimination of the over-parameterized partial models. In the case of the AIC criterion the factor $\gamma(n_{\mathcal{V}}, n_p)$ is equal $2n_p$ what leads to the following final form of the criterion:

$$Q_{\mathrm{AIC}}(\hat{y}_{n,k}^{(l)}) = n_{\mathcal{V}} \log J_{n_{\mathcal{V}}}(N_{\mathrm{arch}}) + 2n_p. \tag{2.90}$$

In the case of the FPE criterion the statistics reflect an expected variance of prediction error during prediction of new observations based on the model obtained for the identification data set:

$$Q_{\mathrm{FPE}}(\hat{y}_{n,k}^{(l)}) = \mathcal{E}(s_{\tau}^2(\tau, N_{\mathrm{arch}})), \tag{2.91}$$

where τ denotes the prediction period. In [23] it is shown that the statistics (2.91) can be approximated by the following expression:

$$Q_{\mathrm{FPE}}(\hat{y}_{n,k}^{(l)}) \approx \Lambda(1 + n_p/n_{\mathcal{V}}), \tag{2.92}$$

where an asymptotic unbiased estimate of the Λ is:

$$\hat{\Lambda} = \frac{J_{n_{\mathcal{V}}}(N_{\mathrm{arch}})}{(1 - n_p/n_{\mathcal{V}})}. \tag{2.93}$$

As a result of substituting (2.93) into (2.92) the final form of the FPE criterion is obtained:

$$Q_{\mathrm{FPE}}(\hat{y}_{n,k}^{(l)}) = J_{n_{\mathcal{V}}}(N_{\mathrm{arch}}) \frac{1 + n_p/n_{\mathcal{V}}}{1 - n_p/n_{\mathcal{V}}}. \tag{2.94}$$

In the case of the AIC criterion, it is possible to select better partial model based on the inequality defined with the statistics (2.89):

$$n_{\mathcal{V}} \log J_{n_{\mathcal{V}}}(N_{\mathrm{arch},1}) + \gamma(n_{\mathcal{V}}, n_{p,1}) \leq n_{\mathcal{V}} \log J_{n_{\mathcal{V}}}(N_{\mathrm{arch},2}) + \gamma(n_{\mathcal{V}}, n_{p,2}), \tag{2.95}$$

where after simple transformation it has the form:

$$J_{n_{\mathcal{V}}}(N_{\mathrm{arch},1}) \leq J_{n_{\mathcal{V}}}(N_{\mathrm{arch},2}) \exp\left[\frac{(\gamma(n_{\mathcal{V}}, n_{p,2}) - \gamma(n_{\mathcal{V}}, n_{p,1}))}{n_{\mathcal{V}}}\right], \tag{2.96}$$

and finally:

$$\chi^2_{\alpha}(n_{p,2} - n_{p,1}) = n_{\mathcal{V}} \left(\exp\left[\frac{(\gamma(n_{\mathcal{V}}, n_{p,2}) - \gamma(n_{\mathcal{V}}, n_{p,1}))}{n_{\mathcal{V}}}\right] - 1\right). \tag{2.97}$$

Based on (2.97) the AIC criterion can be perceived as the F-test [23] with a defined confidence level. The same disadvantage occurs in the case of the FPE criterion. In [23], it is theoretically and practically proved that for $n_{p,2} - n_{p,1} = 1$ degree of freedom, the confidence level is 0.157. This result means that the probability of the selection of over-parameterized structure $N_{\mathrm{arch},2}$ via the AIC or FPE criteria is 15.7%. If the number of partial models in the GMDH neural network is higher than the probability of selection over-parameterized neurons is not acceptable. Another reason, opposite to the application of the AIC and FPE criteria, is the fact that the probability of the selection over-parameterized partial neurons does not decrease along with $n_{\mathcal{V}} \to \infty$. Furthermore, the AIC and FPE criteria are designed for the comparison of the hierarchical partial models $N_{\mathrm{arch},1} \subset N_{\mathrm{arch},2}$. In this case the application of the AIC and FPE evaluation criteria during synthesis of the GMDH network may result in the increase of the neural model uncertainty. For this reason it is necessary to define a new evaluation criteria which will take into account the elementary model uncertainty. This kind of the evaluation criteria is proposed in Sect. 5.3.

2.6.2 Parameters Estimate Inaccuracy

In Sect. 2.6.1 the contributions of the neural network structure errors to the model uncertainty are presented. Apart from the model structure selection stage, an inaccuracy of the parameters estimate also contribute to the model uncertainty. Most of them result from the application of the LMS algorithm to the parameters estimation. Indeed, while applying this approach to the parameters estimation of partial models (2.40), a set of restrictive assumptions has to be satisfied. One of them concerns the transformation with $f^{-1}(\cdot)$. Let us consider the following system output signal:

$$y_k = f\left(\left(\boldsymbol{r}_{n,k}^{(l)}\right)^T \boldsymbol{p}_n^{(l)}\right) + \varepsilon_{n,k}^{(l)} \tag{2.98}$$

The use of linear-in-parameters estimation methods for the model (2.40), e.g., the LMS requires to transform the output of the system (2.98) as follows:

$$\left(\boldsymbol{r}_{n,k}^{(l)}\right)^T \boldsymbol{p}_n^{(l)} = f^{-1}\left(y_k\right) - \varepsilon_{n,k}^{(l)}. \tag{2.99}$$

Unfortunately, the transformation of (2.98) with $f^{-1}(\cdot)$ results in

$$\left(\boldsymbol{r}_{n,k}^{(l)}\right)^T \boldsymbol{p}_n^{(l)} = f^{-1}\left(y_k - \varepsilon_{n,k}^{(l)}\right). \tag{2.100}$$

The noise $\varepsilon_{n,k}^{(l)}$ in (2.99) does not equal to noise in (2.100). Thus, good results can be only expected when the noise $\varepsilon_{n,k}^{(l)}$ magnitude is relatively small.

Another assumption is directly connected with the properties of the LMS. In order to attain the estimator $\hat{\boldsymbol{p}}_{n,k}^{(l)}$ of $\boldsymbol{p}_{n,k}^{(l)}$ for (2.40), which is unbiased and minimum variance [21], it has to be assumed that noise $\boldsymbol{\varepsilon}$ which affects on the system output y_k has the following properties:

$$\mathcal{E}\left[\boldsymbol{\varepsilon}_n^{(l)}\right] = 0, \tag{2.101}$$

$$\operatorname{cov}\left[\boldsymbol{\varepsilon}_n^{(l)}\right] = \left(\sigma_n^{(l)}\right)^2 \boldsymbol{I}. \tag{2.102}$$

The above presented assumptions concerning the properties of the noise may have been one of the potential sources of model uncertainty. The assumption (2.101) means that there are no deterministic disturbances, which unfortunately are usually caused by the structural errors. Moreover, the condition (2.102) means that the model uncertainty is described in a purely stochastic way (uncorrelated noise). It has to be underlined that the assumptions (2.101) and (2.102) are not usually fulfilled in practice which cause an increase of the model uncertainty.

Let us suppose that in some cases the conditions (2.101) and (2.102) are satisfied. Then, it can be shown that $\hat{\boldsymbol{p}}_{n,k}^{(l)}$ (the parameters estimate vector for a neuron of the first layer) is unbiased and minimum variance [21]. Consequently, the neuron output in the first layer becomes the input to other neurons in the second layer. The neuron output can be described by:

$$\hat{\boldsymbol{y}}_n^{(l)} = \boldsymbol{R}_n^{(l)} \left[\left(\boldsymbol{R}_n^{(l)} \right)^T \boldsymbol{R}_n^{(l)} \right]^{-1} \left(\boldsymbol{R}_n^{(l)} \right)^T \boldsymbol{y}, \tag{2.103}$$

where $\boldsymbol{R}_n^{(l)} = [\boldsymbol{r}_n^{(l)}(1), \ldots, \boldsymbol{r}_n^{(l)}(n_T)]^T$, n_T stands for the number of the input-output measurements, $\boldsymbol{y} = [y_1, \ldots, y_{n_T}]^T$ and $\hat{\boldsymbol{y}}_n^{(l)} = [\hat{y}_{n,1}^{(l)}, \ldots, \hat{y}_{n,n_T}^{(l)}]^T$ represents system output vector and its estimate. Apart from the situation in the first layer ($l = 1$), where the matrix $\boldsymbol{R}_n^{(l)}$ depends on $\boldsymbol{u}$, in the subsequent layers $\boldsymbol{R}_n^{(l+1)}$ depends on (2.103) and hence:

$$\mathcal{E} \left[\left[\left(\boldsymbol{R}_n^{(l+1)} \right)^T \boldsymbol{R}_n^{(l+1)} \right]^{-1} \left(\boldsymbol{R}_n^{(l+1)} \right)^T \boldsymbol{\varepsilon}_n^{(l+1)} \right] \neq 0. \tag{2.104}$$

That is why the parameters estimator in the next layers is biased and no there is minimum variance, i.e.

$$\begin{aligned} \mathcal{E}\left[\hat{\boldsymbol{p}}_n^{(l+1)}\right] &= \mathcal{E} \left[\left[\left(\boldsymbol{R}_n^{(l+1)} \right)^T \boldsymbol{R}_n^{(l+1)} \right]^{-1} \left(\boldsymbol{R}_n^{(l+1)} \right)^T \boldsymbol{y} \right] = \\ &= \mathcal{E} \left[\left[\left(\boldsymbol{R}_n^{(l+1)} \right)^T \boldsymbol{R}_n^{(l+1)} \right]^{-1} \left(\boldsymbol{R}_n^{(l+1)} \right)^T \right. \\ &\quad \left. \left(\boldsymbol{R}_n^{(l+1)} \boldsymbol{p}_n^{(l+1)} + \boldsymbol{\varepsilon}_n^{(l+1)} \right) \right] = \\ &= \boldsymbol{p}_n^{(l+1)} + \mathcal{E} \left[\left[\left(\boldsymbol{R}_n^{(l+1)} \right)^T \boldsymbol{R}_n^{(l+1)} \right]^{-1} \left(\boldsymbol{R}_n^{(l+1)} \right)^T \boldsymbol{\varepsilon}_n^{(l+1)} \right]. \end{aligned} \tag{2.105}$$

To settle this problem, the instrumental variable method or other methods mentioned in [16] can be employed. On the other hand, these methods provide only asymptotic convergence, and hence a large data set is usually required to obtain an unbiased parameters estimate.

All mentioned in Sects. 2.6.1 and 2.6.2 sources of model uncertainty cause that it is important to modify the process of the synthesis of the GMDH network in order to reduce its uncertainty. Moreover, it is necessary to develop the methodology of the calculation of the model uncertainty allowing to design the scheme of robust FDI.

2.7 Concluding Remarks

In the present chapter a few architectures of neural networks are presented which can be applied for the identification of dynamic non-linear systems. In particular, the globally and locally recurrent dynamic neural networks are described. The advantages and disadvantages of these models are widely described. One of the most important factors influencing the quality of the neural model is an appropriate selection of the neural model architecture during system identification. In this chapter the Pruning, Bottom-up and Discrete optimisation methods, which are commonly applied in practice, are described. Unfortunately, the effectiveness of the Top-down methods usually depends on the initial conditions, learning process, and order of training data samples. The obtained with the application of such methods neural models cannot be assumed as optimal in the sense of number of neurons or parameters. In the Bottom-up methods the termination condition of the network synthesis is connected to the quality criterion applied during the network training. For this reason neural networks with poor generalisation properties can be obtained. These limitations are not present in the case of discrete optimisation methods. However, such methods have significant computational cost requirements.

In order to solve such problems the methodology of designing of neural models with the application of the GMDH approach is proposed. According to appropriate partial models structure selection, proper evaluation criteria and selection methods choosing the models with good quality and generalization properties can be obtained. Additional advantage of the GMDH neural networks is that the realization of dynamic at the level of each partial model is performed. This strategy simplifies the training process of the network. Moreover, the stability of the each neuron and the whole GMDH network can be relatively easily ensured. The application of the GMDH approach can improve the quality of the model but cannot eliminate the uncertainty of the model at all. It should be noticed that apart from the contribution of the structure errors to the model uncertainty the parameters estimates inaccuracy also influences the model quality. The knowledge about the sources of the GMDH model uncertainty allows to improve the synthesis process of such a neural model. Moreover, the mathematical description of the model uncertainty enables to develop the robust fault detection schemes.

3

Estimation Methods in Training of ANNs for Robust Fault Diagnosis

3.1 Introduction

The classical ANNs training problem is formulated as the parameters estimation of non-linear-in-parameter models. Among the existing algorithms, a few groups can be distinguished: gradient-based algorithms, evolutionary algorithms and stochastic algorithms [25, 26, 29, 31]. Undoubtedly, the gradient-based algorithms form the most popular group of them. Their properties can be summarized as follows: low computational burden, easy implementation and fast performance. On the other hand, the training of the ANN is usually an optimisation problem of a multimodal cost function. This means that the gradient-based algorithms usually find one of the unsatisfactory local minima. To overcome this problem, it seems desirable to use either stochastic or evolutionary algorithms [128, 129]. It is well known that such algorithms possess global convergence properties. Unfortunately, the number of parameters of an ANN is rather large and it leads to an increase of computational cost which makes the parameters estimation process time consuming.

The application of the GMDH approach [69, 70, 117] during the neural network synthesis allows to apply the parameters estimation algorithms for linear-in-parameters models, e.g., the LMS. It follows from the fact that in the case of the GMDH neural network the problem of the network training boils down to the estimation of parameters of single neurons. The advantage of this approach is a simple computation algorithm that gives good results even for small sets of measuring date. Unfortunately, the usual statistical parameters estimation framework assumes that the data are corrupted by the errors, which can be modeled as the realizations of independent random variables with a known or parameterized distribution. A more realistic approach is to assume that the errors lie between given prior bounds. It leads directly to the bounded error set estimation class of algorithms, e.g., the BEA [16, 130],

M. Mrugalski, *Advanced Neural Network-Based Computational Schemes for Robust Fault Diagnosis*, Studies in Computational Intelligence 510,
DOI: 10.1007/978-3-319-01547-7_3,

OBE [15, 131] and ZBA [132, 133, 134], which can be employed to solve the parameters estimation problem.

These algorithms also enable to perform the robust parameters estimation based on the fault diagnosis. The idea of this method is based on the parameters estimation of the diagnosed system and its comparison with a'priori estimated parameters of the nominal system. It should be noticed that this strategy allows to perform both the fault detection and isolation.

The chapter is organized as follows: Section 3.2 presents the concept of the robust FDI with the application of the parameters estimation methods. Section 3.3 is devoted to methods for parameters and its uncertainty estimation. In particular, the LMS, BEA, OBE and ZBA are described. Section 3.4 contains an illustrative example devoted to the effectiveness comparison of the early presented parameters estimation methods. Finally, in Sect. 3.5 an example of the application of the FDI of a brushed DC motor is included. It should be also pointed out that the results described in this chapter are based on [135, 136].

3.2 Robust FDI via Parameters Estimation

The residual-based fault detection approaches are widely applied in practice. Unfortunately, the main drawback of these kind of techniques is difficulty with isolation of the faulty parameters which causes the fault of the diagnosed system. On the other hand, the parameter-based fault diagnosis scheme is very often applied in industrial practice. Unfortunately, most of the applied methods rely on the measurement of the parameters of the diagnosed system $\boldsymbol{p}$ and comparison of their values with the nominal parameters $\boldsymbol{p}_n$. If the absolute value of the difference between these values is negligible small, it is assumed that the diagnosed system is healthy:

$$|\boldsymbol{p}_n - \boldsymbol{p}| < \delta_p. \tag{3.1}$$

The weakness of these methods is the limited range of practical application because the measurements of the parameters of the diagnosed systems are often not available. In order to solve this problem the methods of the parameters estimation on the basis of the diagnosed system inputs and outputs can be applied. This kind of approaches can be only used when the relation between parameters and system inputs and outputs is known. Unfortunately, the effectiveness of this fault diagnosis scheme depends on the quality of the parameters estimate. If the obtained parameters estimate is biased then the false alarms of undetected faults can appear. This situation can appear because the data sets used during the parameters estimation are often corrupted by the measurements noise or disturbances. In order to overcome these disadvantages an appropriate parameters estimates techniques which take the noises and disturbances into their account have to be used.

The idea of the proposed approach relies on the application of the parameters estimation methods to parameters estimation with its uncertainty based on the input and output $\{\boldsymbol{r}_k, y_k\}$ of the diagnosed system (cf. Fig. 3.1).

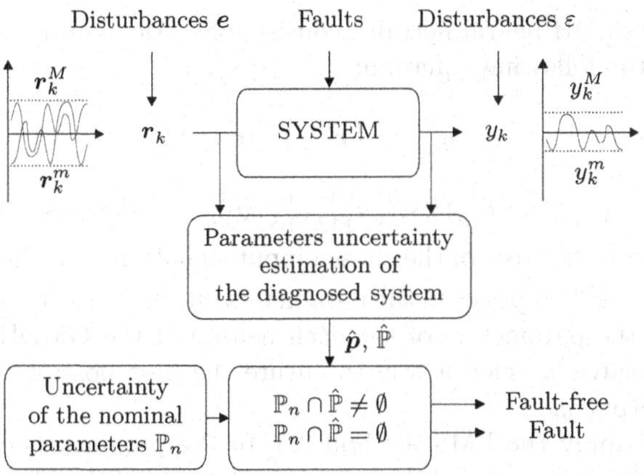

Fig. 3.1. Scheme of the robust FDI via parameters estimation

The comparison of the estimated parameters and its uncertainty with the a'priori estimated parameters of the nominal system allows to perform both the faults detection and faults isolation (cf. Fig. 3.2).

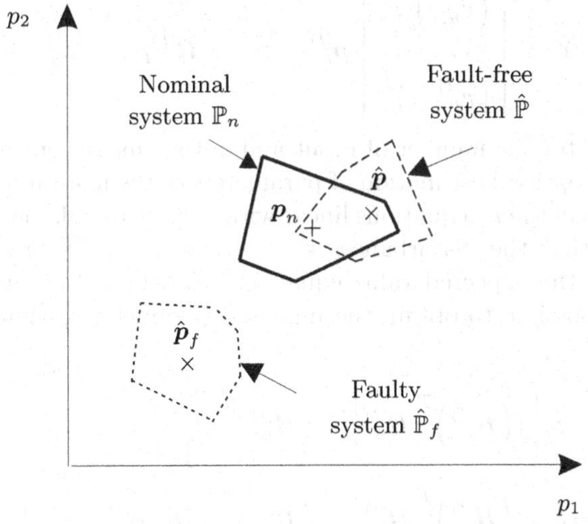

Fig. 3.2. Parameters estimation in the robust FDI

3.3 Techniques of Parameters and Uncertainty Estimation

3.3.1 Least-Mean Square Method

The classical GMDH neural network consists of static neurons which output is defined by the following equation:

$$y_{n,k}^{(l)} = \left(\boldsymbol{r}_{n,k}^{(l)}\right)^T \hat{\boldsymbol{p}}_n^{(l)}, \tag{3.2}$$

where $\boldsymbol{r}_{n,k}^{(l)} = [(u_{i,k}^{(l)})^2, (u_{j,k}^{(l)})^2, u_{i,k}^{(l)}u_{j,k}^{(l)}, u_{i,k}^{(l)}, u_{j,k}^{(l)}, 1]^T$ denotes the regression vector created on the basis of the chosen input signals $\boldsymbol{u}_k^{(l)}$ for the l-th GMDH layer, and $\hat{\boldsymbol{p}}_n^{(l)} \in \mathbb{R}^{n_p}$ represents the parameters estimate. As it was mentioned in Sect. (2.4) the parameters of the each neuron of the GMDH neural network are estimated in such a way to ensure the best possible modelling of the system output y_k.

In order to apply the LMS method [23] to the parameters estimation of the neuron (3.2), it is assumed that the output signal of the identified system is written as:

$$y_k = \left(\boldsymbol{r}_{n,k}^{(l)}\right)^T \boldsymbol{p}_n^{(l)} + \varepsilon_{n,k}^{(l)}, \tag{3.3}$$

where $\boldsymbol{p}_n^{(l)} \in \mathbb{R}^{n_p}$ denotes the real unknown values of the parameters, $\varepsilon_{n,k}^{(l)}$ represents disturbances and $\boldsymbol{r}_{n,k}^{(l)}$ is the known regressor vector. The relation (3.3) can be written for the whole measurement data set in the following form:

$$\boldsymbol{Y} = \begin{bmatrix} \left(\boldsymbol{r}_{n,1}^{(l)}\right)^T \\ \dots \\ \left(\boldsymbol{r}_{n,n_{\mathcal{D}}}^{(l)}\right)^T \end{bmatrix} \boldsymbol{p}_n^{(l)} + \boldsymbol{\varepsilon}_n^{(l)} = \boldsymbol{R}_n^{(l)}\boldsymbol{p}_n^{(l)} + \boldsymbol{\varepsilon}_n^{(l)}, \tag{3.4}$$

where $n_{\mathcal{D}}$ denotes the number of input and output measurements of data.

The task of optimal estimation of parameters of the neuron (3.2), relies on the calculation of the n_p equations linear with respect to unknown parameters $\boldsymbol{p}$. Assuming that the disturbances $\boldsymbol{\varepsilon} = [\varepsilon_1, \varepsilon_2, \dots, \varepsilon_{n_{\mathcal{D}}}]^T$ are independent variables with the expected value equal zero (2.101) and covariance matrix (2.102), it is possible to obtain the unbiased parameters estimate (minimal variance):

$$\begin{aligned}\mathcal{E}\left[\hat{\boldsymbol{p}}_n^{(l)}\right] &= \mathcal{E}\left[\left[\left(\boldsymbol{R}_n^{(l)}\right)^T \boldsymbol{R}_n^{(l)}\right]^{-1} \left(\boldsymbol{R}_n^{(l)}\right)^T \boldsymbol{y}\right] = \\ &= \mathcal{E}\left[\left[\left(\boldsymbol{R}_n^{(l)}\right)^T \boldsymbol{R}_n^{(l)}\right]^{-1} \left(\boldsymbol{R}_n^{(l)}\right)^T \left(\boldsymbol{R}_n^{(l)}\boldsymbol{p}_n^{(l)} + \boldsymbol{\varepsilon}_n^{(l)}\right)\right] = \\ &= \boldsymbol{p}_n^{(l)} + \left[\left(\boldsymbol{R}_n^{(l)}\right)^T \boldsymbol{R}_n^{(l)}\right]^{-1} \left(\boldsymbol{R}_n^{(l)}\right)^T \mathcal{E}\left[\boldsymbol{\varepsilon}_n^{(l)}\right] = \boldsymbol{p}_n^{(l)},\end{aligned} \tag{3.5}$$

for which the covariance matrix is:

$$\operatorname{cov}\left[\hat{\boldsymbol{p}}_n^{(l)}\right] = \mathcal{E}\left[\left[\hat{\boldsymbol{p}}_n^{(l)} - \mathcal{E}\left[\hat{\boldsymbol{p}}_n^{(l)}\right]\right]\left[\hat{\boldsymbol{p}}_n^{(l)} - \mathcal{E}\left[\hat{\boldsymbol{p}}_n^{(l)}\right]\right]^T\right] = \\ = \left(\sigma_n^{(l)}\right)^2 \left[\left(\boldsymbol{R}_n^{(l)}\right)^T \boldsymbol{R}_n^{(l)}\right]^{-1}, \tag{3.6}$$

which is described by:

$$\hat{\boldsymbol{p}}_n^{(l)} = \left[\left(\boldsymbol{R}_n^{(l)}\right)^T \boldsymbol{R}_n^{(l)}\right]^{-1} \left(\boldsymbol{R}_n^{(l)}\right)^T \boldsymbol{Y}. \tag{3.7}$$

In the case of the GMDH neural network consists of the dynamic neurons with IIR filters (2.59), which are presented in Sect. 2.5.1 it is possible to describe the neuron model in the analogical form as (3.2) whereas the regressor vector $\boldsymbol{r}_{n,k}^{(l)}$ consists of $\boldsymbol{r}_k = [-y_{n,k-1}^{(l)}, \ldots, -y_{n,k-n_a}^{(l)}, \boldsymbol{u}_{n,k}^{(l)}, \boldsymbol{u}_{n,k-1}^{(l)}, \ldots, \boldsymbol{u}_{n,k-n_b}^{(l)}]$ and the parameters estimates are $\hat{\boldsymbol{p}}_n^{(l)} = [a_1, \ldots, a_{n_a}, v_0, v_1, \ldots, v_{n_b}]$. Similarly to the static model, it is possible to obtain the parameters estimate using the classical LMS method. Unfortunately, the parameters estimate (3.7) is not linear in the $\boldsymbol{Y}$, what follows from the fact that the regressor depends on the components of the vector $\boldsymbol{Y}$ according to $\boldsymbol{r}_k$. These properties result from the fact that the regression matrix consisting regression vectors for the subsequent measurements:

$$\boldsymbol{R}_n^{(l)} = \begin{bmatrix} \left(\boldsymbol{R}_{n,1}^{(l)}\right)^T \\ \cdots \\ \left(\boldsymbol{R}_{n,n_D}^{(l)}\right)^T \end{bmatrix}, \tag{3.8}$$

is correlated with the previous disturbances influencing the outputs. In spite of the fulfilled condition (2.101) but taking into account the correlation of the matrices $\boldsymbol{R}_n^{(l)}$ and $\varepsilon_n^{(l)}$ it is obtained:

$$\left[\left(\boldsymbol{R}_n^{(l)}\right)^T \boldsymbol{R}_n^{(l)}\right]^{-1} \left(\boldsymbol{R}_n^{(l)}\right)^T \mathcal{E}\left[\varepsilon_n^{(l)}\right] \neq 0. \tag{3.9}$$

Finally, the obtained parameters estimate is unbiased:

$$\mathcal{E}\left[\hat{\boldsymbol{p}}_n^{(l)}\right] = \boldsymbol{p}_n^{(l)} + \left[\left(\boldsymbol{R}_n^{(l)}\right)^T \boldsymbol{R}_n^{(l)}\right]^{-1} \left(\boldsymbol{R}_n^{(l)}\right)^T \mathcal{E}\left[\varepsilon_n^{(l)}\right] \neq \boldsymbol{p}_n^{(l)}. \tag{3.10}$$

Knowing the correlation and characteristic of the disturbances it is possible to modify the LMS algorithm in order to obtain the unbiased parameters estimate [23]. Unfortunately, this kind of estimation methods has a narrow scope of applications.

The instrumental variable estimation is an alternative approach applied to the parameters estimation of the dynamical models [137]. This approach is based on the assumption that the additional measurement data set is available in the form of the matrix $\boldsymbol{Z}$. The matrix of instrumental variable has the dimensions and structure similar to the matrix $\boldsymbol{R}$. The measurement data in the matrix $\boldsymbol{Z}$ is asymptotically correlated with the data included in the matrix $\boldsymbol{R}$ and they are asymptotically uncorrelated with disturbances:

$$P = \lim_{n_{\mathcal{D}} \to \infty} \left[\frac{1}{n_{\mathcal{D}}} \left(\boldsymbol{Z}_n^{(l)} \right)^T \boldsymbol{\varepsilon}_n^{(l)} \right] = \mathbf{0}. \tag{3.11}$$

Moreover, the existence of the non-singular matrix is required:

$$P = \lim_{n_{\mathcal{D}} \to \infty} \left[\frac{1}{n_{\mathcal{D}}} \left(\boldsymbol{Z}_n^{(l)} \right)^T \boldsymbol{R}_n^{(l)} \right] = \boldsymbol{R}_{\mathbf{zx}}. \tag{3.12}$$

After the fulfilment of the above assumptions, the parameters estimator of the instrumental variables for the dynamical system (2.59) receives the following form:

$$\hat{\boldsymbol{p}}_n^{(l)} = \left[\left(\boldsymbol{Z}_n^{(l)} \right)^T \boldsymbol{R}_n^{(l)} \right]^{-1} \left(\boldsymbol{Z}_n^{(l)} \right)^T \mathbf{Y}. \tag{3.13}$$

Moreover, the estimator is asymptotically unbiased:

$$\begin{aligned}
\hat{\boldsymbol{p}}_n^{(l)} &= \left[\left(\boldsymbol{Z}_n^{(l)} \right)^T \boldsymbol{R}_n^{(l)} \right]^{-1} \left(\boldsymbol{Z}_n^{(l)} \right)^T \boldsymbol{y} = \\
&= \left[\left(\boldsymbol{Z}_n^{(l)} \right)^T \boldsymbol{R}_n^{(l)} \right]^{-1} \left(\boldsymbol{Z}_n^{(l)} \right)^T \left(\boldsymbol{R}_n^{(l)} \boldsymbol{p}_n^{(l)} + \boldsymbol{\varepsilon}_n^{(l)} \right) = \\
&= \left[\left(\boldsymbol{Z}_n^{(l)} \right)^T \boldsymbol{R}_n^{(l)} \right]^{-1} \left(\boldsymbol{Z}_n^{(l)} \right)^T \boldsymbol{R}_n^{(l)} \boldsymbol{p}_n^{(l)} + \\
&+ \left[\left(\boldsymbol{Z}_n^{(l)} \right)^T \boldsymbol{R}_n^{(l)} \right]^{-1} \left(\boldsymbol{Z}_n^{(l)} \right)^T \boldsymbol{\varepsilon}_n^{(l)} = \\
&= \boldsymbol{p}_n^{(l)} + \left[\left(\boldsymbol{Z}_n^{(l)} \right)^T \boldsymbol{R}_n^{(l)} \right]^{-1} \left(\boldsymbol{Z}_n^{(l)} \right)^T \boldsymbol{\varepsilon}_n^{(l)} = \\
&= \boldsymbol{p}_n^{(l)} + \left[\frac{1}{n_{\mathcal{D}}} \left(\boldsymbol{Z}_n^{(l)} \right)^T \boldsymbol{R}_n^{(l)} \right]^{-1} \left[\frac{1}{n_{\mathcal{D}}} \left(\boldsymbol{Z}_n^{(l)} \right)^T \boldsymbol{\varepsilon}_n^{(l)} \right],
\end{aligned} \tag{3.14}$$

what can be proved with the application of (3.11) and (3.12):

$$P = \lim_{n_{\mathcal{D}} \to \infty} \hat{\boldsymbol{p}}_n^{(l)} = \boldsymbol{p}_n^{(l)} + \boldsymbol{R}_{\mathbf{zx}}^{-1} \mathbf{0} = \boldsymbol{p}_n^{(l)}. \tag{3.15}$$

On the basis of (3.15) it can be noticed that the instrumental variable estimation is not a perfect method because the obtained parameters estimate is biased for a small number of the measurement data samples $n_{\mathcal{D}}$. Moreover, the additional disadvantage of this method is that the results of the estimation depend on the choice of the instrumental variable matrix $\boldsymbol{Z}$ [138].

3.3.2 Bounded-Error Approach

The usual statistical parameters estimation framework assumes that the data are corrupted by errors which can be modelled as the realisations of independent random variables with a known or parameterised distribution. A more realistic approach is to assume that the errors lie between given prior bounds. This is the case, for example, for the data collected with an analogue-to-digital converter or for measurements performed with a sensor of a given type. Such reasoning leads directly to the BEA [15, 16, 130, 139, 140]. Let us consider the following system:

$$y_k = \left(\boldsymbol{r}_{n,k}^{(l)}\right)^T \boldsymbol{p}_n^{(l)} + \varepsilon_{n,k}^{(l)}. \tag{3.16}$$

The problem is to obtain the parameters estimate vector $\hat{\boldsymbol{p}}_{n,k}^{(l)}$, as well as an associated parameter uncertainty required to design robust fault detection system. In order to simplify the notation, the index ${}_n^{(l)}$ is omitted. The knowledge regarding the set of admissible parameter values allows to obtain the confidence region of the model output which satisfies:

$$\hat{y}_k^m \leq y_k \leq \hat{y}_k^M, \tag{3.17}$$

where $\hat{y}_k^m$ and $\hat{y}_k^M$ are the minimum and maximum admissible values of the model output that are consistent with the input-output measurements of the system. Under the assumptions described in Sect. 2.6, the uncertainty of the neural network can be obtained according to [141].

It is assumed that ε_k is bounded as follows:

$$\varepsilon_k^m \leq \varepsilon_k \leq \varepsilon_k^M, \tag{3.18}$$

where the bounds ε_k^m and ε_k^M are known a'priori. The idea underlying the BEA is to obtain a feasible parameters set [15]. This set can be defined as

$$\mathbb{P} = \left\{\boldsymbol{p} \in \mathbb{R}^{n_p} \mid y_k - \varepsilon_k^M \leq \boldsymbol{r}_k^T \boldsymbol{p} \leq y_k - \varepsilon_k^m \,, k = 1, \ldots, n_{\mathcal{T}}\right\}, \tag{3.19}$$

where $n_{\mathcal{T}}$ is the number of input-output measurements. It can be perceived as a region of parameter space that is determined by $n_{\mathcal{T}}$ pairs of hyperplanes:

$$\mathbb{P} = \bigcap_{k}^{n_{\mathcal{T}}} \mathbb{S}(k), \tag{3.20}$$

where each pair defines the parameter strip:

$$\mathbb{S}(k) = \left\{\boldsymbol{p} \in \mathbb{R}^{n_p} \mid y_k - \varepsilon_k^M \leq \boldsymbol{r}_k^T \boldsymbol{p} \leq y_k - \varepsilon_k^m\right\}. \tag{3.21}$$

Any parameter vector contained in $\mathbb{P}$ is a valid estimate of $\boldsymbol{p}$. In practice, the centre (in a geometrical sense) of $\mathbb{P}$ (cf. Fig. 3.3 for $n_p = 2$) is chosen as the parameters estimate $\hat{\boldsymbol{p}}$, e.g.:

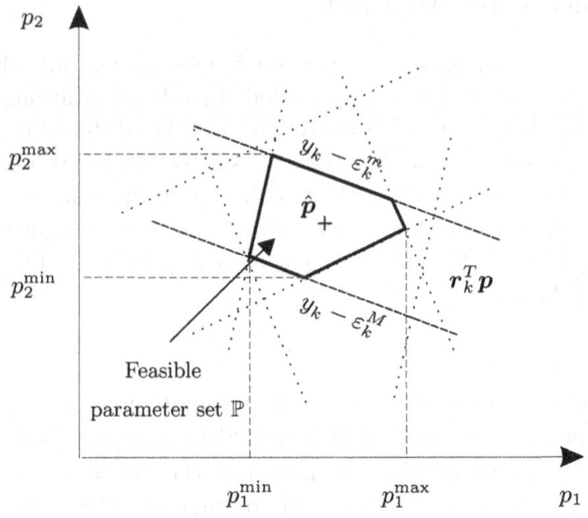

Fig. 3.3. Feasible parameters set for the error-free regressor obtained via the BEA

$$\hat{p}_i = \frac{p_i^{\min} + p_i^{\max}}{2}, \quad i = 1, \ldots, n_p, \tag{3.22}$$

where:

$$p_i^{\min} = \arg\min_{p \in \mathbb{P}} p_i, \quad i = 1, \ldots, n_p, \tag{3.23}$$

$$p_i^{\max} = \arg\max_{p \in \mathbb{P}} p_i, \quad i = 1, \ldots, n_p. \tag{3.24}$$

The problem (3.23) and (3.24) can be solved with well-known linear programming techniques [15, 142], but when $n_{\mathcal{T}}$ and/or n_p are large, computational cost may be significant. It constitutes the main drawback of the approach. One way out of this problem is to apply a technique where constraints are executed separately one after another [143]. Although this approach does not constitute a perfect remedy for the computational problem being considered. This means that the described BEA can be employed for the tasks with a relatively small dimension as it is in the case for the GMDH neurons. The main difficulty associated with the BEA concerns a'priori knowledge regarding the error bounds ε_k^m and ε_k^M. However, these bounds can be also estimated [15] by assuming that $\varepsilon_k^m = \varepsilon^m$, $\varepsilon_k^M = \varepsilon^M$ and then suitably extending the unknown parameter vector $\boldsymbol{p}$ by ε^m and ε^M. Another assumption is as follows: $\varepsilon^m = -\varepsilon^{h*}$ and $\varepsilon^M = \varepsilon^{h*}$. The task of determining the bounds can now be formulated:

$$(\varepsilon^m, \varepsilon^M) = \arg \min_{\varepsilon^M \geq 0,\ \varepsilon^m \leq 0} \varepsilon^M - \varepsilon^m, \tag{3.25}$$

with respect to the following constraints:

$$y_k - \varepsilon^M \leq \boldsymbol{r}_k^T \boldsymbol{p} \leq y_k - \varepsilon^m \ , k = 1, \ldots, n_{\mathcal{T}}. \tag{3.26}$$

The main disadvantage of the BEA is high computational burden. It follows from a complex shape of the polytope $\mathbb{P}$ resulting from the intersection of the $n_{\mathcal{U}}$ hyperplanes $\mathbb{S}$. The increasing number of measurements $n_{\mathcal{U}}$ allows to obtain the exact shape of the multidimensional polytope $\mathbb{P}$, and hence more accurate estimate of parameters. Unfortunately, the subsequent intersections of the feasible parameters set $\mathbb{P}$ by the hyperplanes $\mathbb{S}$ increase the number of the vertices and surfaces of more and more complex polytope $\mathbb{P}$. The number of vertices $\varpi(n_s, n_p)$ of the polytope $\mathbb{P}$ depends on the number of surfaces of polytope n_s and dimension of the unknown parameters n_p:

$$\varpi(n_s, n_p) = \begin{cases} \sum_{j=1}^{m} = \frac{n_s}{n_s - j} \binom{n_s - j}{j} \binom{j}{j - n_p} & \text{for} \quad n_p = 2m, \\ \sum_{j=0}^{m} = \frac{n_p + 1}{n_s - j} \binom{n_s - j}{j} \binom{j}{j - n_p} & \text{for} \quad n_p = 2m + 1. \end{cases} \tag{3.27}$$

In order to overcome the problem of high computational cost a wide class of algorithms approximating (cf. Fig. 3.4) of the exact polytope $\mathbb{P}$ by the ellipsoid, orthotopic, zonotopic or paralleltopic bounding have been developed [15, 132, 144, 145, 146].

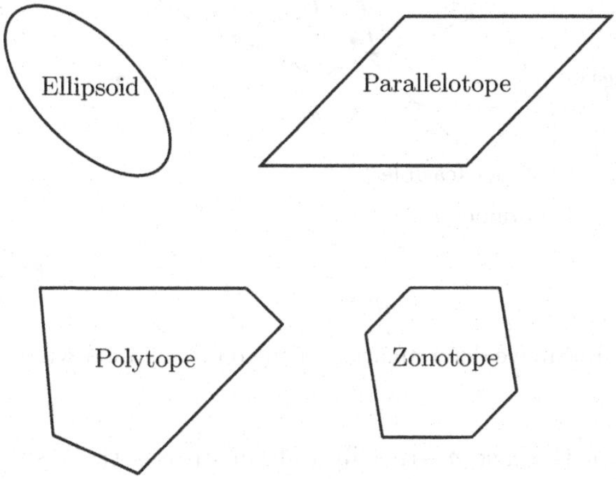

Fig. 3.4. Feasible parameters sets obtained via approximation algorithms

3.3.3 Outer Bounding Ellipsoid Algorithms

The OBE algorithms constitute the most numerous group of approaches applied to the parameters estimation by the approximation of the exact feasible parameters set [15, 131, 147, 148, 149]. The concept of the OBE algorithm (cf. Fig. 3.5) relies on the approximation of the convex polytopes $\mathbb{P}_k$ by simpler ellipsoids $\mathbb{E}_k$:

$$\mathbb{E}_k \subset \mathbb{P}_k, \tag{3.28}$$

In this recursive algorithm the feasible parameters set is obtained by the intersection of the strip $\mathbb{S}_k$ and the ellipsoid $\mathbb{E}_{k-1}$ (3.21) and its outerbanding by a new $\mathbb{E}_k$ ellipsoid for all subsequent $k = 1, ..., n_{\mathcal{D}}$ data points:

$$\mathbb{E}_k \supset \mathbb{E}_{k-1} \cap \mathbb{S}_k. \tag{3.29}$$

In this way the measurements of inputs and outputs represented in the form of the strips $\mathbb{S}_k$ are taken into account one after the another to construct a succession of ellipsoids containing all values of $\boldsymbol{p}$ consistent with all previous measurements. The OBE algorithm provides rules for computing $\boldsymbol{p}_k$ and $\boldsymbol{P}_k$ in such a way that the volume of $\mathbb{E}(\hat{\boldsymbol{p}}_k, \boldsymbol{P}_k)$ is minimized (cf. Fig. 3.5).

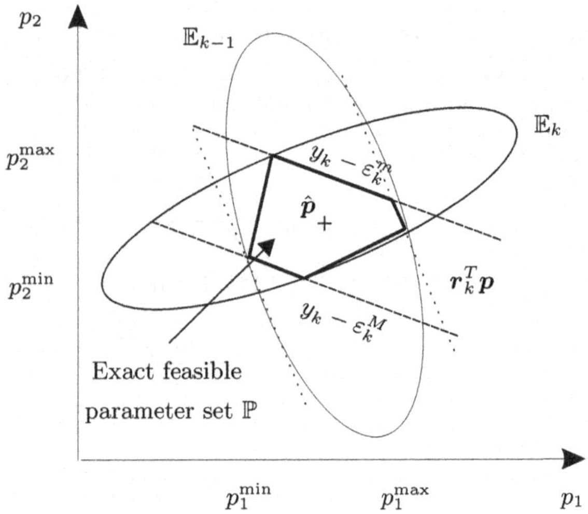

Fig. 3.5. Recursive determination of the outer ellipsoid with the OBE

The center of the last $n_{\mathcal{T}}$-th ellipsoid constitutes the resulting parameters estimate while the ellipsoid itself represents the feasible parameters set. However, any parameter vector $\hat{\boldsymbol{p}}$ contained in $\mathbb{E}(n_{\mathcal{T}})$ is a valid estimate of $\boldsymbol{p}$.

The initial ellipsoid containing the expected values of the parameters can be defined as follows:

$$\mathbb{E}_0 = \left\{ \boldsymbol{p} \in \mathbb{R}^{n_p} \,|(\boldsymbol{p} - \hat{\boldsymbol{p}}_0)^T \boldsymbol{P}_0^{-1} (\boldsymbol{p} - \hat{\boldsymbol{p}}_0) \leq \sigma_0^2, \boldsymbol{P}_0 \sigma_0^2 = \frac{1}{\tau} \mathrm{i}(n_p) \right\}, \tag{3.30}$$

where: τ represents some small value that $\mathbb{E}_0$ contain $\mathbb{P}_k$ for all $k = 1, \ldots, n_p$, $\boldsymbol{P}_0$ and σ_0^2 represents a'priori knowledge about the identified system. The subsequent ellipsoids are defined:

$$\begin{aligned}\mathbb{E}_k =&\{\boldsymbol{p} \in \mathbb{R}^{n_p} \,|\alpha_k(\boldsymbol{p} - \hat{\boldsymbol{p}}_{k-1})^T \boldsymbol{P}_{k-1}^{-1}(\boldsymbol{p} - \hat{\boldsymbol{p}}_{k-1})+ \\ &\beta_k(y_k - \boldsymbol{r}_k^T \boldsymbol{p})^2 \le \alpha_k \sigma_{k-1}^2 + \beta_k \delta_k^2\},\end{aligned} \tag{3.31}$$

where $\alpha_k \in [0, 1]$ is the forgetting factor which weights the information from the previous data, while $\beta_k \in [0, 1]$ denotes the selecting factor which weights new data. The center of the k-th ellipsoid, which volume and orientation are defined by the matrix $\boldsymbol{P}_k$, represents k-th estimate of the parameters. The relation (3.31) can be rewritten as follows:

$$\mathbb{E}_k =\{\boldsymbol{p} \in \mathbb{R}^{n_p} \,|(\boldsymbol{p} - \hat{\boldsymbol{p}}_k)^T \boldsymbol{P}_k^{-1}(\boldsymbol{p} - \hat{\boldsymbol{p}}_k) \le \sigma_k^2\}, \tag{3.32}$$

where the parameters $\hat{\boldsymbol{p}}_k$, $\boldsymbol{P}_k$ and σ_k^2 are calculated according to the following equations:

$$v_k = y_k - \boldsymbol{r}_k^T \hat{\boldsymbol{p}}_{k-1}, \tag{3.33}$$

$$\mathcal{G}_k = \boldsymbol{r}_k^T \boldsymbol{P}_{k-1} \boldsymbol{r}_k, \tag{3.34}$$

$$\boldsymbol{P}_k = \frac{1}{\alpha_k}\left[\boldsymbol{P}_{k-1} - \frac{\beta_k \boldsymbol{P}_{k-1}\boldsymbol{r}_k \boldsymbol{r}_k^T \boldsymbol{P}_{k-1}}{\alpha_k + \beta_k \mathcal{G}_k}\right], \tag{3.35}$$

$$\hat{\boldsymbol{p}}_k = \hat{\boldsymbol{p}}_{k-1} + \beta_k \boldsymbol{P}_k \boldsymbol{r}_k v_k, \tag{3.36}$$

$$\sigma_k^2 = \alpha_k \sigma_{k-1}^2 + \beta_k \delta_k^2 - \frac{\alpha_k \beta_k v_k^2}{\alpha_k + \beta_k \mathcal{G}_k}. \tag{3.37}$$

Depending on the choice of values of the parameters α_k and β_k the two main classes of the OBE algorithm can be distinguished. The first group of algorithm is based on the minimization of the geometrical size of the ellipsoid $\mathbb{E}_k$. In this methods the parameters α_k and β_k are calculated to minimize the scalar measure of the matrix $\boldsymbol{P}_k$ which describes the geometrical size of the ellipsoid $\mathbb{E}_k$. Equations (3.33-3.37) are applied and additionally, it is assumed that $\alpha_k = 1/\sigma_{k-1}^2$ and $\beta_k = \lambda_k/\delta_k^2$. The variable λ_k, which value depends on the measurements data, is obtained as a result of the minimization of the measure reflecting the geometrical size of the ellipsoid. In [148] two measures are minimized:

$$\mu_{d,k} = \det(\delta_k^2 \boldsymbol{P}_k), \tag{3.38}$$

and

$$\mu_{t,k} = \mathrm{trace}(\delta_k^2 \boldsymbol{P}_k). \tag{3.39}$$

These measures reflect the volume and sum of squares of semi-axes of the ellipsoid $\mathbb{E}_k$. The algorithm using eq. (3.33-3.37), in which λ_k is calculated on the basis of the measure (3.38), is called the Minimal-Volume OBE algorithm. In this approach λ_k constitutes the solution of the equation:

$$a_1 \lambda_k^2 + a_2 \lambda_k + a_3 = 0, \tag{3.40}$$

where:

$$a_1 = (n_{\mathcal{U}} - 1)\sigma_{k-1}^4 \mathcal{G}_k 2, \tag{3.41}$$

$$a_2 = \left((2n_{\mathcal{U}} - 1)\delta_k^2 - \sigma_{k-1}^2 \mathcal{G}_k + v_k^2\right) \sigma_{k-1}^2 \mathcal{G}_k, \tag{3.42}$$

$$a_3 = \left((n_{\mathcal{U}}(\delta_k^2 - v_k) - \sigma_{k-1}^2 \mathcal{G}_k\right) \delta_k^2. \tag{3.43}$$

The optimal value of λ_k is provided by:

$$\lambda_k = \begin{cases} 0 & \text{when} \quad a_3 \geq 0 \\ \left(-a_2 + \left(a_2^2 - 4a_1a_3\right)^{\frac{1}{2}}\right)/2a_1, & \text{otherwise.} \end{cases} \tag{3.44}$$

The algorithm based on the measure (3.39) to calculation of λ_k is called the Minimal Trace OBE algorithm and the value of λ_k is the solution of equation:

$$\lambda_k^3 + b_1\lambda_k^2 + b_2\lambda_k + b_3 = 0, \tag{3.45}$$

where:

$$b_1 = \frac{3\delta_k^2}{\sigma_{k-1}^2 \mathcal{G}_k}, \tag{3.46}$$

$$b_2 = \frac{\delta_k^2 \mathcal{G}_k \left[\mu_{k-1}\left(\delta_k^2 - v_k^2\right) - \sigma_{k-1}^4 \gamma_k\right]}{\psi_k} + \frac{2\delta_k^2\left[\delta_k^2 \mathcal{G}_k \mu_{k-1} - \sigma_{k-1}^2 \gamma_k \left(\delta_k^2 - v_k^2\right)\right]}{\psi_k}, \tag{3.47}$$

$$b_3 = \frac{\delta_k^4\left[\left(\delta_k^2 - v_k^2\right)\mu_{k-1} - \sigma_{k-1}^4 \gamma_k\right]}{\left(\sigma_{k-1}^2 \psi_k\right)}, \tag{3.48}$$

$$\gamma_k = \sigma_k^T \boldsymbol{P}_{k-1}^2 \sigma_k, \tag{3.49}$$

$$\psi_k = \sigma_{k-1}^4 \mathcal{G}_k 2\left[\mathcal{G}_k \mu_{k-1} - \sigma_{k-1}^2 \gamma_k\right]. \tag{3.50}$$

The optimal value λ_k is then given by:

$$\lambda_k = \begin{cases} 0 & \text{when} \quad b_3 \geq 0 \\ \lambda_k^\star & \text{otherwise} \end{cases} \tag{3.51}$$

where $\lambda_k^\star$ is the positive real root of eq. (3.45).

The Degenerate Minimal Volume OBE algorithm represents the next class of OBE algorithms. These algorithms minimize the value σ_k^2 describing the size of the ellipsoid, or the sequence of values $\{\sigma_k^2\}$ is non-increasing, i.e., $\sigma_k^2 \leq \sigma_{k-1}^2$. In the algorithm proposed by [150] it is assumed that $\alpha_k = 1 - \lambda_k$ and $\beta_k = \lambda_k$, where λ_k value is the solution of the following constrained minimization problem:

$$\min_{\lambda_k} \sigma_k^2, \quad \text{for} \quad 0 \leq \lambda_k \leq \tau < 1, \tag{3.52}$$

where the designed variable $\tau \in [0, 1]$ has to ensure that the matrix $\boldsymbol{P}_k$ is bounded. The value of λ_k is calculated according to the following equations:

$$\lambda_k = \begin{cases} 0 & \text{when} \quad \gamma_k \geq 1 \\ \min(\tau, \varrho_k) & \text{for} \quad 0 < \tau < 1 \text{ otherwise}, \end{cases} \tag{3.53}$$

where:

$$\gamma_k = \frac{\delta_k^2 - \sigma_{k-1}^2}{\upsilon_k^2}, \tag{3.54}$$

$$\varrho_k = \begin{cases} \tau & \text{when} \quad \upsilon_k = 0 \\ \frac{(1-\gamma_k)}{2} & \text{when} \quad \mathcal{G}_k = 1 \\ \tau & \text{when } \gamma_k(\mathcal{G}_k - 1) + 1 \leq 0 \\ \frac{1}{1-\mathcal{G}_k}\left(1 - \left(\frac{\mathcal{G}_k}{\gamma_k(\mathcal{G}_k-1)+1}\right)^{\frac{1}{2}}\right) & \text{when } \gamma_k(\mathcal{G}_k - 1) + 1 > 0. \end{cases} \tag{3.55}$$

In the next degenerated Minimal Volume OBE algorithm it is assumed that the factor α_k is constant and equals to λ_k, when $\beta_k = \lambda_k$. The value λ_k is calculated to minimize the criterion (3.52), however without taking into account the constraint $0 \leq \lambda_k \leq \tau$. After substitution $\alpha_k = \lambda$ and $\beta_k = \lambda_k$ to (3.33-3.37), allowing to obtain the values of parameters estimate $\hat{\boldsymbol{p}}_k$, uncertainty matrix $\boldsymbol{P}_k$ and σ_k^2 have the following form:

$$\hat{\boldsymbol{p}}_k = \hat{\boldsymbol{p}}_{k-1} + \frac{\lambda_k \boldsymbol{P}_{k-1}\boldsymbol{r}_k \upsilon_k}{\lambda + \lambda_k \mathcal{G}_k}, \tag{3.56}$$

$$\boldsymbol{P}_k = \frac{1}{\lambda}\left[\boldsymbol{P}_{k-1} - \frac{\lambda_k \boldsymbol{P}_{k-1}\boldsymbol{r}_k \boldsymbol{r}_k^T \boldsymbol{P}_{k-1}}{\lambda + \lambda_k \mathcal{G}_k}\right], \tag{3.57}$$

$$\sigma_k^2 = \lambda\sigma_{k-1}^2 + \lambda_k \delta_k^2 - \frac{\lambda\lambda_k \upsilon_k^2}{\lambda + \lambda_k \mathcal{G}_k}. \tag{3.58}$$

The values λ_k, which minimize σ_k^2, can be calculated as follows:

$$\lambda_k = \begin{cases} 0 & \text{when} \quad |\upsilon_k| \leq \delta_k \quad \text{or} \quad \mathcal{G}_k = 0 \\ \frac{\lambda}{\mathcal{G}_k}\left(\frac{|\upsilon_k|}{\delta_k} - 1\right), & \text{otherwise}. \end{cases} \tag{3.59}$$

3.3.4 Zonotope-Based Algorithm

The main drawback of the BEA is associated computational cost and the complexity of the representation of the exact feasible parameters set. In order to overcome these limitations the approaches based on the approximation of the exact feasible parameters set can be applied. As it was mentioned in the introduction several overbounding approaches can be used in order to obtain this goal. In this subsection the approach based on the application of the zonotopes is presented [132, 133, 134]. In this algorithm, similarly to earlier presented estimation techniques, the parameterized form of the identified system is assumed:

$$y_k = \boldsymbol{r}_k^T \boldsymbol{p} + \varepsilon_k. \tag{3.60}$$

However, in this algorithm the other form of the bounding of the errors ε_k, which represents the modeling uncertainty and the measurements noise, is assumed:

$$\varepsilon_k \in \mathbb{R} : |\varepsilon| \leq \sigma. \tag{3.61}$$

On the beginning, on the basis of the $k = 1, ..., n_T$ given inputs $\boldsymbol{r}_k$ and outputs y_k measurements, the data strips bounding the consistent parameters are defined:

$$\mathbb{S}_k = \{\boldsymbol{p} : -\sigma \leq \boldsymbol{r}_k^T \boldsymbol{p} - y_k \leq \sigma\}. \tag{3.62}$$

In the next step, the zonotope of m order which approximates the feasible parameters set representing the parameters uncertainty has to be defined as:

$$\mathbb{Z}_{m,k} = \hat{\boldsymbol{p}} \oplus \mathbb{P}\boldsymbol{B}^m = \{\hat{\boldsymbol{p}} + \mathbb{P}z : z \in \boldsymbol{B}^m\}, \tag{3.63}$$

where:

- $\hat{\boldsymbol{p}} \in \mathbb{R}^{n_p}$ – given vector of parameters,
- $\mathbb{P} \in \mathbb{R}^{n_p \times m}$ – matrix of parameters uncertainty,
- $\boldsymbol{B}^m \in \mathbb{R}^{m \times 1}$ – the unitary box composed of m unitary intervals $\boldsymbol{B} = [-1, 1]$,
- $\oplus$ – the Minkowski sum of two sets (e.g., $\mathcal{X} \oplus \mathcal{Y} = \{x + y : x \in \mathcal{X}, y \in \mathcal{Y}\}$).

In the fact the zonotope of m order is defined as the Minkowski sum of segments defined by the columns of matrix $\mathbb{P} : \hat{\boldsymbol{p}} \oplus \mathbb{P}\boldsymbol{B}^m = \hat{\boldsymbol{p}} \oplus \mathbb{P}_1\boldsymbol{B} \oplus \mathbb{P}_2\boldsymbol{B} \oplus, ..., \oplus \mathbb{P}_m\boldsymbol{B}$ where $\mathbb{P}_i$ denotes the i-th column of matrix $\mathbb{P}$ and $\mathbb{P}_i\boldsymbol{B} = \{\mathbb{P}_i b : b \in [-1, 1]\}$. It should be underlined that if $m < n_p$ or $\mathbb{P}$ is not a full-rank matrix then the zonotope $\mathbb{Z}_m$ has zero volume. Moreover, the zonotope is always a bounded set provided that the entries of matrix $\mathbb{P}$ are bounded. Furthermore, the order of zonotope m represents its geometrical complexity (cf. Fig. 3.6).

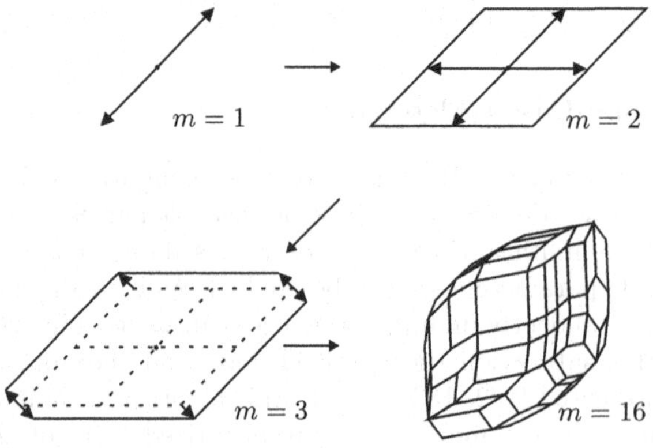

Fig. 3.6. Zonotope construction for $m = 1$, $m = 2$, $m = 3$ and $m = 15$

The intersection of the defined zonotope $\mathbb{Z}_{m,k}$ with the subsequent data strips $\mathbb{S}_k$ allows to obtain the approximation feasible solution set containing the parameters estimates:

$$AFSS_{k+1} = \mathbb{S}_k \cap \mathbb{Z}_k, \quad \text{for} \quad k = 1, ..., n_d. \tag{3.64}$$

Unfortunately, the application of the above identification algorithm can lead to increase of the zonotope complexity for each subsequent sample time k. For this reason, it is necessary to use an algorithm of overbounding of the higher-order zonotope by the lower-order zonotope. In order to achieve this goal, it is advisable to define so-called zonotope support strip $\mathbb{S}_{s,k}$ for a given zonotope $\mathbb{Z}_{m,k}$ and the vector $\boldsymbol{r}_k \in \mathbb{R}^{n_r}$:

$$\mathbb{S}_{s,k} = \{\boldsymbol{p} : q_b \leq \boldsymbol{r}_k^T \boldsymbol{p} \leq q_a\}, \tag{3.65}$$

where q_a and q_b satisfy:

$$q_a = \max_{\boldsymbol{p} \in \mathbb{Z}} \boldsymbol{r}_k^T \boldsymbol{p}, \tag{3.66}$$

and

$$q_b = \min_{\boldsymbol{p} \in \mathbb{Z}} \boldsymbol{r}_k^T \boldsymbol{p}. \tag{3.67}$$

Knowing that $\mathbb{Z}_{m,k} = \hat{\boldsymbol{p}} \oplus \mathbb{P}\boldsymbol{B}^m$ and the vector $\boldsymbol{r}_k \in \mathbb{R}^{n_r}$, the values of q_a and q_b can be calculated according to the following equations:

$$q_a = \boldsymbol{r}_k^T \hat{\boldsymbol{p}} + \|\mathbb{P}^T \boldsymbol{r}_k\|_1, \tag{3.68}$$

and:

$$q_b = \boldsymbol{r}_k^T \hat{\boldsymbol{p}} - \|\mathbb{P}^T \boldsymbol{r}_k\|_1, \tag{3.69}$$

where $\|\mathbb{P}^T \boldsymbol{r}_k\|_1$ represents the sum of the absolute values of the components of $\mathbb{P}^T \boldsymbol{r}_k$. Moreover, for a given zonotope $\mathbb{Z}_{m,k} = \hat{\boldsymbol{p}} \oplus \mathbb{P}\boldsymbol{B}^m$ and strip $\mathbb{S}_k$, the zonotope tight strip $\mathbb{S}_{t,k}$ is obtained as:

$$\mathbb{S}_{t,k} = \mathbb{S}_k \cap \mathbb{S}_{s,k}. \tag{3.70}$$

The intersection of the tight strips $\mathbb{S}_{t,k} = \{\boldsymbol{p} : |\boldsymbol{r}_k^T \boldsymbol{p} - y_k| \leq \sigma\}$ and the zonotope $\mathbb{Z}_{m,k}$ leads to obtaining a new reduced order zonotope $\mathbb{Z}_{r,k} = \hat{\boldsymbol{p}} \oplus \mathbb{P}\boldsymbol{B}^r = \hat{\boldsymbol{p}} \oplus [\mathbb{P}_1, \mathbb{P}_2, ..., \mathbb{P}_r]\boldsymbol{B}^r$:

$$\mathbb{Z}_{r,k} = \mathbb{Z}_{m,k} \cap \mathbb{S}_{t,k} \subseteq \hat{\boldsymbol{p}}_j \oplus T_j \boldsymbol{B}^r, \tag{3.71}$$

where $\hat{\boldsymbol{p}}_j$ and T_j for every integer j where $0 \leq j \leq r$ can be calculated according the following equations:

$$\hat{\boldsymbol{p}}_j = \begin{cases} \hat{\boldsymbol{p}} + \left(\frac{y_k - \boldsymbol{r}_k^T \hat{\boldsymbol{p}}}{\boldsymbol{r}_k^T \mathbb{P}_j}\right)\mathbb{P}_j, & \text{if} \quad 1 \leq j \leq r \quad \text{and} \quad \boldsymbol{r}_k^T \mathbb{P}_j \neq 0 \\ \hat{\boldsymbol{p}}, & \text{otherwise,} \end{cases} \tag{3.72}$$

$$T_j = \begin{cases} [T_1^j, T_2^j, ..., T_r^j], & \text{if} \quad 1 \leq j \leq r \quad \text{and} \quad \boldsymbol{r}_k^T \mathbb{P}_j \neq 0 \\ \mathbb{P}, & \text{otherwise}, \end{cases} \tag{3.73}$$

$$T_i^j = \begin{cases} \mathbb{P}_i - (\frac{\boldsymbol{r}_k^T \mathbb{P}_i}{\boldsymbol{r}_k^T \mathbb{P}_j})\mathbb{P}_j, & \text{if} \quad i \neq j \\ (\frac{\sigma}{\boldsymbol{r}_k^T \mathbb{P}_j})\mathbb{P}_j, & \text{if} \quad i = j. \end{cases} \tag{3.74}$$

In order to minimize the volume of a new zonotope $\mathbb{Z}_{r,k}$ bounding the intersection, it is necessary to chose the optimal value of j_{opt} from $r+1$ possible choices of j:

$$j_{opt} = \arg \min_{0 \leq j \leq r} \text{vol}(\hat{\boldsymbol{p}}_j \oplus T_j \boldsymbol{B}^r), \tag{3.75}$$

and in order to make this task less computationally demanding, the choice of j_{opt} can be done by:

$$j_{opt} = \arg \min_{0 \leq j \leq r} \det(T_j T_j^T). \tag{3.76}$$

3.4 An Illustrative Example – LMS vs. BEA, OBE, ZBA

The purpose of the present section is to show the effectiveness of the proposed approaches based on the LMS, BEA, OBE and ZBA in the task of the parameters estimation of the neurons in the GMDH network. Let us consider the following static system:

$$y_k = p_1 \sin(u_k^2) + p_2 u_k + \varepsilon_k,$$

where the nominal values of the parameters are $\boldsymbol{p}_n = [1.5, 0.5]^T$, the input data u_k and the noise ε_k, $k = 1, \ldots, n_T$, are generated according to the uniform distribution, i.e., $u_k \in \mathcal{U}(0,2)$ and $\varepsilon_k \in \mathcal{U}(-0.05, 0.1)$, where $\mathcal{U}(\cdot)$ stands for the uniform distribution. It should be noticed that the noise ε_k does not satisfy (2.101). The problem is to obtain the parameters estimate $\hat{\boldsymbol{p}}$ and the corresponding feasible parameters set using the set of input-output measurements $\{u_k, y_k)\}_{k=1}^{n_T=200}$. In order to obtain the feasible parameters set for the LMS method, the F-test [21] is employed and a 95% confidence region is obtained, whilst in the case of the BEA that the initial boundary values of the disturbances $\varepsilon_k^m = -0.1$, $\varepsilon_k^M = 0.2$ are assumed.

To tackle this task, the approaches described in Sects. 3.3.1-3.3.4 are employed. In the case of the LMS method the parameters uncertainty is obtained with the confidence level of 95%, whereas the application of the BEA, OBE and ZBA allows to calculate the parameters uncertainty with the confidence level of 100%. Parameters estimates and their minimal and maximal values of the parameters estimates for all these methods are presented in Tab. 3.1. Figures 3.7-3.10 show the parameters estimates and the feasible parameters sets obtained with the LMS, BEA, OBE and ZBA algorithms, respectively.

Table 3.1. Parameters estimates obtained with the application of the LMS, BEA, OBE and ZBA algorithms

$\hat{\boldsymbol{p}}$	$\hat{p}_1$	$[\hat{p}_1^{\min}, \hat{p}_1^{\max}]$	$\hat{p}_2$	$[\hat{p}_2^{\min}, \hat{p}_2^{\max}]$
LMS	1.5170	[1.4803, 1.5537]	0.5265	[0.5068, 0.5461]
BEA	1.5004	[1.4992, 1.5016]	0.5001	[0.4995, 0.5008]
OBE	1.5007	[1.4473, 1.5541]	0.4925	[0.4649, 0.5201]
ZBA	1.5111	[1.4751, 1.5472]	0.5153	[0.4980, 0.5326]

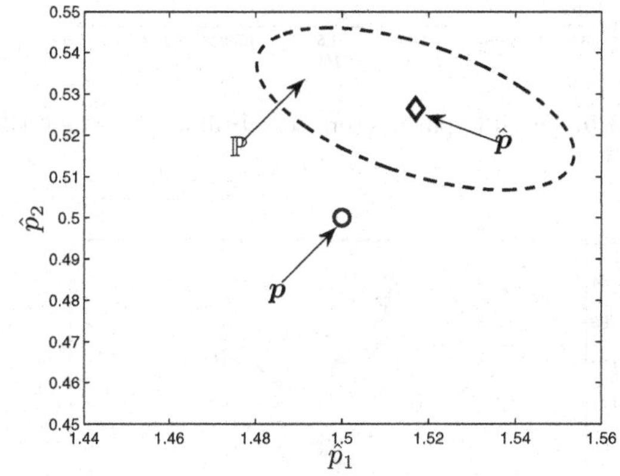

Fig. 3.7. Feasible parameters set obtained via the LMS

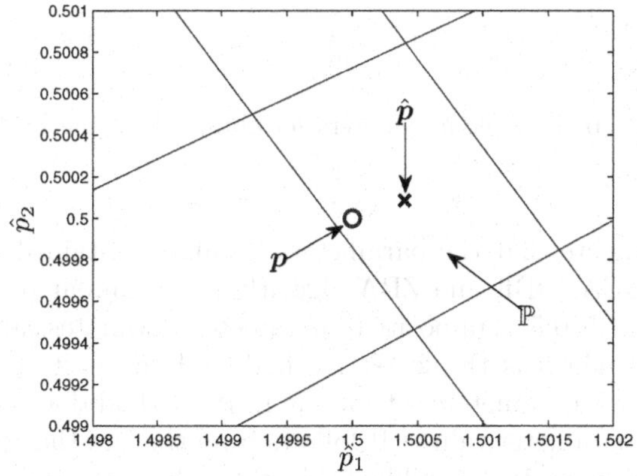

Fig. 3.8. Feasible parameters set obtained via the BEA

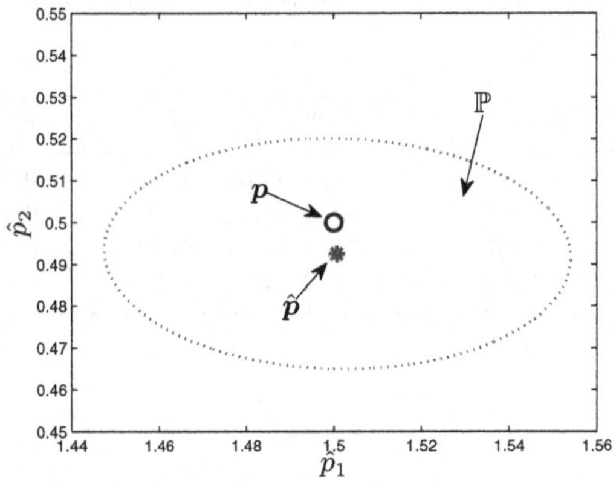

Fig. 3.9. Feasible parameters set obtained via the OBE

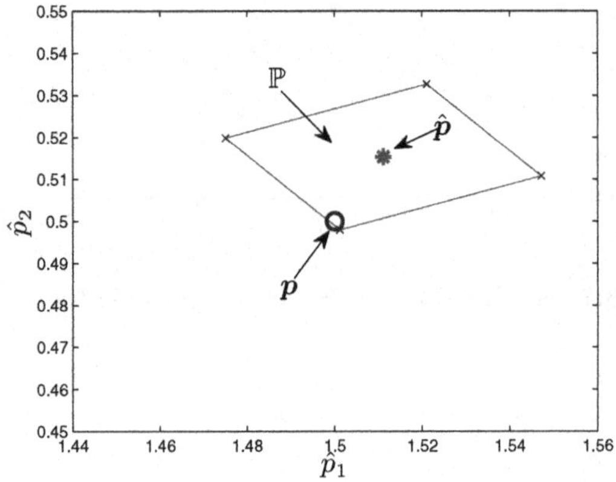

Fig. 3.10. Feasible parameters set obtained via the ZBA

The results indicate that the parameters estimates obtained with the application of the BEA, OBE and ZBA algorithms are similar to the nominal parameters $\boldsymbol{p}_n = [1.5, 0.5]^T$, opposite to parameters estimates calculated with the LMS method which is the worst method. Indeed, as it can be seen in Fig. 3.7, the feasible parameters set for the LMS method does not even contain the nominal parameter values. It follows from the fact that the condition (2.101) concerning noise is not fulfilled. Moreover, from the results presented in Fig. 3.8 it is clear that the BEA is superior to the other methods. This method provides the parameters estimate which is the closest to the nominal parameters values. Furthermore, the obtained feasible parameters set has the smallest size. Unfortunately, this accuracy is occupied by a large computational cost. The parameters estimates obtained with the application of the OBE and ZBA algorithms are quite accurate, however, they are not so good as those obtained with the application of the BEA. It should be emphasized that the feasible parameters set obtained with the OBE algorithm is quite large (cf. Fig. 3.9). In the case of the GMDH neural network, that is composed of many neurons, such uncertainty can be accumulated and hence, the performance of the network can be significantly degraded. From this reason, the ZBA seem to be attractive trade-off between the accuracy offered by the BEA and performance given by the OBE algorithm.

3.5 Robust FDI of Brushed DC Motor

The objective of this section is to design a fault detection scheme based on the approaches described in the previous sections. For the modeling and fault diagnosis purpose a Maxon DC motor $RE25 - 10W118743$ presented in Fig. 3.11 was chosen [151]. It consists of two pole permanent magnet and precious metal brushes. This brushed motor has no magnetic cogging and the ironless winding. The motor has a high acceleration thanks to a low inertia. Due to a eleventh segmented commutator there is only a small torque ripple. The winding has very specific advantages which cause that there is no magnetic detent and minimal electromagnetic interference. The efficiency of the motor is up to 90%. The nominal voltage is $12V$ and no load current $26mA$. The model of the brushed DC motor implemented in the Matlab Simulink [151], is used to generate the data sets applied to modelling purpose. In this model, two subsystems can be differentiated: electrical and mechanical. The electrical subsystem is derived as follows:

Fig. 3.11. Brushed Maxon DC motor RE25-10W 118743

$$i_k = \frac{1}{R}u_k + \frac{k_e}{R}\omega_{r,k}, \tag{3.77}$$

$$\omega_{r,k} = \frac{k_t}{J}i_k - \exp^{(\frac{-Tc_v}{J})}\omega_{r,k-1}. \tag{3.78}$$

The vector of the input variables of the electrical subsystem includes voltage and rotor angular velocity $\boldsymbol{r}_{e,k} = [u_k, \omega_{r,k}]$, whereas the electrical subsystem response is the current i_k. In the case of the mechanical subsystem, the vector of input variables consists of current and delayed rotor angular velocity $\boldsymbol{r}_{m,k} = [i_k, \omega_{r,k-1}]$. The mechanical subsystem response is the rotor angular velocity $\omega_{r,k}$. The nominal parameters of the DC motor have the values:

- $R = 2.18\ \Omega$ – coil resistance,
- $J = 1.07e-6\ kgm^2$ – rotor inertia,
- $k_t = 0.0235\ Nm/A$ – torque constant,
- $k_e = 0.0235\ Vs/rad$ – back-emf constant,
- $c_v = 12e-7$ – viscose friction constant
- $T = 0.001\ s$ – sampling period.

On the beginning the identification of parameters of the electrical and mechanical subsystems is made on the basis of the fault-free data $\{u_k, i_k, \omega_{r,k}\}_{k=1}^{n_\mathcal{T}=100}$ generated with the application of the nominal values of the motor parameters implemented in Matlab. Furthermore, for the fault diagnosis purpose, the data containing two faults in the electrical and mechanical subsystems are generated. The first fault, in the electrical subsystem, relies on change of back-emf constant from k_e=0.0235Ω to $k_{e,f}$=0.0470Ω. The second fault simulated in the mechanical subsystem, relies on simultaneous increasing of viscose friction coefficient from c_v=12e-7Nms/rad to c_v=18e-7Nms/rad. For the electrical and mechanical subsystems the disturbances $\boldsymbol{e}_k \in U(-0.1, 0.2)$ influencing $\boldsymbol{r}_{e,k}$ and $\boldsymbol{r}_{m,k}$ and disturbances $\varepsilon_k \in U(-0.05, 0.1)$ influencing i_k and $\omega_{r,k}$ are generated according to the uniform distribution. These disturbances represent fluctuation and unbalanced voltage, inaccuracy of control systems and mechanical overload. The LMS, BEA, OBE and ZBA algorithms are further applied (by the assumption that the nominal parameters are not known) to calculate the parameters estimates for the electrical subsystem based only on the data set $\{[u_k, \omega_{r,k}], i_k\}$. Moreover, these algorithms are also applied to calculate the parameters estimates in the mechanical subsystem based only on the data set $\{[i_k, \omega_{r,k-1}], \omega_{r,k}\}$. Table 3.2 shows the estimates obtained by applying all the methods for the fault-free and faulty electrical motor. The results show that the parameters estimates obtained with the application of the BEA, OBE, and ZBA algorithms are similar to the nominal parameters for electrical $\boldsymbol{p}_n = [0.4587, 0.0108]$ and mechanical $\boldsymbol{p}_n = [21962.6, 0.8452]$ subsystem assumed during the simulation of the fault-free motor, opposite to parameters estimates calculated with the LMS method. It follows from the fact that the condition (2.101) concerning noise is not fulfilled. The FDI approaches based on the parameters estimation algorithms are also applied to the parameters and their uncertainty estimation for data containing the faults simulated in the electrical and mechanical subsystems. The results contained in Tab. 3.2 and in Fig. 3.13 and 3.12 clearly show that the parameters estimates are significantly different from the nominal parameters and similar to the faulty parameters in the electrical $\hat{\boldsymbol{p}}_f = [0.4587, 0.0216]$ and mechanical $\hat{\boldsymbol{p}}_f = [21962.6, 0.8452]$ subsystems assumed during the fault simulation.

Table 3.2. Comparison of parameters estimates obtained with the application of the LMS, BEA, OBE and ZBA for the fault-free and faulty electrical motor

$\hat{\boldsymbol{p}}$	$\hat{p}_{n,1}$	$\hat{p}_{n,2}$	$\hat{p}_{f,1}$	$\hat{p}_{f,2}$
Electrical subsystem				
LMS	0.4461	0.0105	0.4614	0.0217
BEA	0.4596	0.0107	0.4562	0.0214
OBE	0.4594	0.0108	0.4504	0.0212
ZBA	0.4632	0.0109	0.4609	0.0217
Mechanical subsystem				
LMS	21152.0	0.9060	19214.0	0.7443
BEA	21965.8	0.8943	21959.2	0.8446
OBE	21953.0	0.8928	21856.0	0.8366
ZBA	21975.0	0.8954	21898.0	0.8426

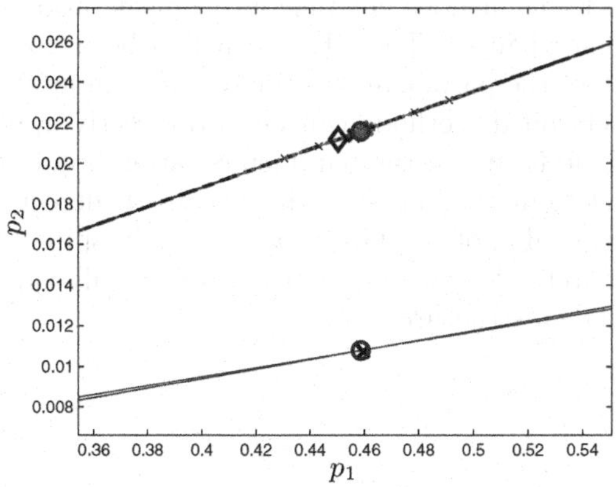

Fig. 3.12. Feasible parameters set obtained for the faulty electrical subsystem

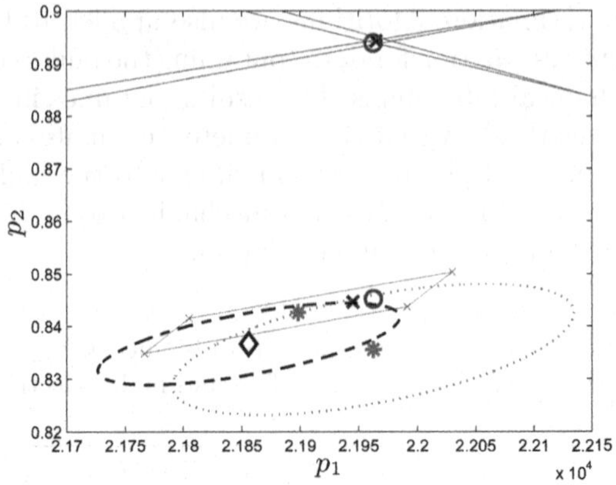

Fig. 3.13. Feasible parameters set obtained for the faulty mechanical subsystem

3.6 Concluding Remarks

The objective of this chapter is to present the algorithms which can be applied to the estimation of neurons parameters. In particular, the LMS, BEA, OBE and ZBA algorithms are described. All these methods allow to estimate the neurons parameters and calculate the neuron uncertainty. This feature is especially important when the robust fault detection system on the basis of neural model is designed. The BEA algorithm ensures the best quality of the parameters estimate and the smallest feasible parameter set. This factor has the main impact on the sensitivity of the robust fault detection system. On the other hand, the LMS method delivers biased parameters estimate what follows from the fact that it does not take into consideration the properties of the noise and disturbances. The ZBA seems to be a good compromise between the quality of the parameters estimate and computational cost.

Moreover, all mentioned methods can be also directly applied to robust fault diagnosis scheme based on the parameters estimation of the diagnosed system. This robust fault diagnosis scheme allows to detect the fault and identify precisely the value of the faulty parameter in spite of disturbances incidence. Unfortunately, this kind of approach can be only applied for simple systems with the known structure.

4

MLP in Robust Fault Detection of Static Non-linear Systems

4.1 Introduction

One of the most desirable features of the model obtained during the system identification is small modelling uncertainty which is defined as a mismatch between the model and the system being considered [10]. It follows from the fact that the effectiveness of the fault detection systems depends on the uncertainty of the neural model and disturbances existing in the industrial system. In the case of the most widely applied the ANNs such as the MLP, the model uncertainty can appear both during structure selection of the neural model and parameters estimation. The contribution of the errors caused by an inappropriate network architecture to model uncertainty follows from the subjective choice of the number of layers and neurons in a particular layer as well as the form of their activation function. This problem can be partly limited by methods which allow an automatic selection of the neural model structure, e.g., the GMDH [116, 152]. Apart from the contribution of the errors caused by the inappropriate structure selection the parameters estimates inaccuracy also influences model uncertainty. It is the result of not taking into account the noises and disturbances during the application of most of the parameters estimation algorithms which may cause the biased parameters estimates.

As the model uncertainty and disturbances are usually difficult to eliminate, there is a need to develop a robust fault detection systems [3, 11, 152, 153]. In order to overcome this problem, it is necessary to obtain a mathematical description of the neural model uncertainty. The solution to this problem can be an application of the OBE algorithm [15, 16]. This approach is based on a realistic assumption that the noises and disturbances lie between given prior bounds. The application of the OBE algorithm allows estimating the parameters of the MLP and the corresponding model uncertainty in the form of the admissible parameter set called also parameter uncertainty. This result allows to define the model uncertainty in the form of the output adaptive threshold, which can be next applied to the robust fault detection.

M. Mrugalski, *Advanced Neural Network-Based Computational Schemes for Robust Fault Diagnosis*, Studies in Computational Intelligence 510,
DOI: 10.1007/978-3-319-01547-7_4, © Springer International Publishing Switzerland 2014

The chapter is organized as follows: Section 4.2 describes the concept of the robust neural model-based fault detection. Section 4.3 introduces the structure of the neural network and the estimation method used during the non-linear system identification. In Sect. 4.4 original developments regarding the uncertainty estimation of the MLP, which is based on the application of the OBE algorithm, are widely described. Section 4.5 contains an illustrative example devoted to the effectiveness comparison of the Non-linear Least-Mean Square (NLMS) and OBE algorithms. Section 4.6 gives a comprehensive study regarding the application of the proposed approach to the robust fault detection of an valve actuator in the Lublin Sugar Factory in Poland. The results presented in this chapter are based on [154, 155] what should be highlighted.

4.2 Robust Model-Based Fault Detection

The most known structure of the fault detection system is based on a model of a diagnosed system. Model-based fault diagnosis can be defined as the detection, isolation and identification of faults in the system based on a comparison of system available measurements with information represented by the system mathematical model (cf. Fig. 4.1) [1, 3]. The model of the diagnosed system is created before its application in the fault diagnosis system.

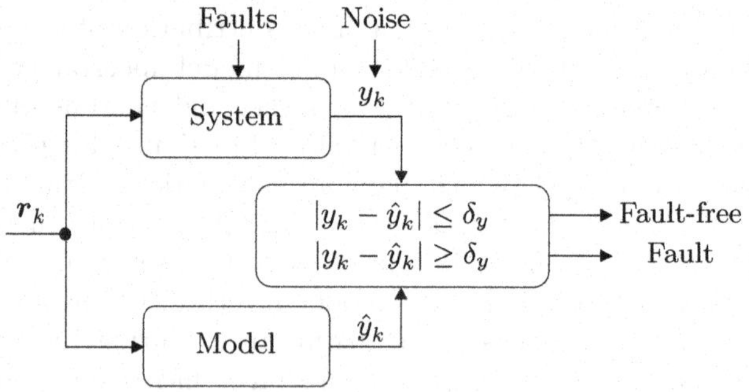

Fig. 4.1. Scheme of the residual-based fault detection system

The comparison of the system y_k and model response $\hat{y}_k$ leads to the generation of the residual ϵ_k:

$$\epsilon_k = y_k - \hat{y}_k, \tag{4.1}$$

which is the source of information about faults for further processing. In the model-based fault detection approach, it is assumed that the residual ϵ_k should normally be close to zero in the fault-free case, and it should be distinguishably different from zero in the case of the fault. In other words, the residual should ideally carry only an information regarding the fault. Under such an assumption, the faults are detected by setting of a constant threshold on the residual signal (cf. Fig. 4.2).

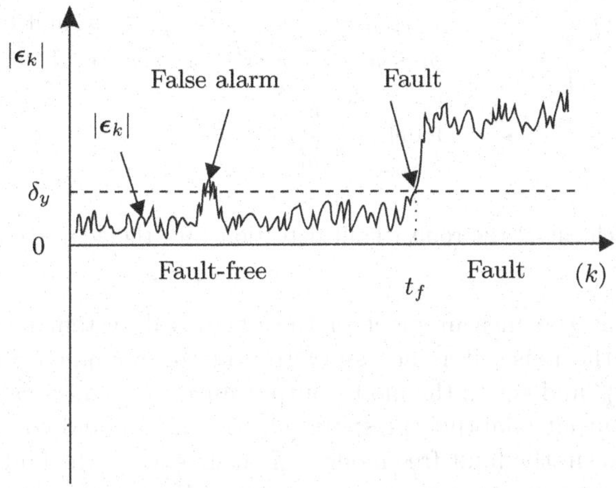

Fig. 4.2. Constant threshold in the fault detection

In this case, the fault can be detected when the absolute value of the residuum $|\epsilon_k|$ will be larger than arbitrary assumed threshold value δ_y:

$$|\epsilon_k| \leq \delta_y. \tag{4.2}$$

The fundamental difficulty with this kind symptom evaluation is that measurement of the system output y_k is usually corrupted by the noise and disturbances $\varepsilon_k^m \leq \varepsilon_k \leq \varepsilon_k^M$, where $\varepsilon_k^m \leq 0$ and $\varepsilon_k^M \geq 0$. Another difficulty follows from the fact that the model obtained during the system identification is usually uncertain [152]. The model uncertainty can appear during model structure selection and also parameters estimation. In practice, due to modelling uncertainty and measurement noise, it is necessary to assign wider thresholds in order to avoid false alarms. It usually implies a reduction of the fault detection sensitivity.

To tackle this problem, the adaptive time-variant threshold, that is adapted according to the system behaviour, can be applied. Indeed, knowing the model structure and possessing the knowledge regarding its uncertainty it is possible to design a robust fault detection scheme. The idea of the proposed approach is illustrated in Fig. 4.3.

The proposed technique relies on the calculation of the model output uncertainty interval based on the estimated parameters which values are known at some confidence level:

$$\hat{y}_k^m \leq \hat{y}_k \leq \hat{y}_k^M. \tag{4.3}$$

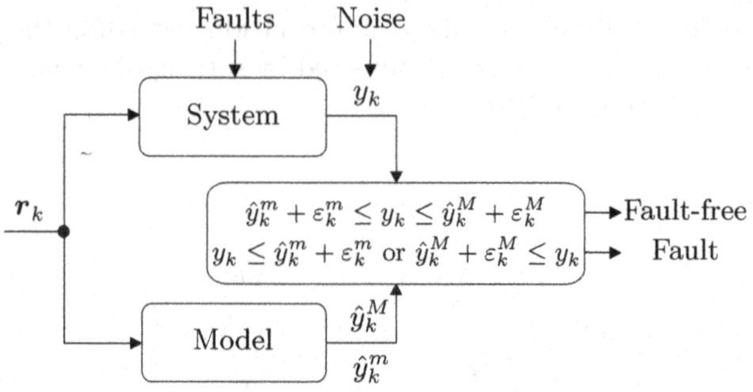

Fig. 4.3. Scheme of the robust fault detection with the adaptive threshold

Additionally, as the measurement of the diagnosed system response y_k is corrupted by the noise, it is necessary to add the boundary values of the output error ε_k^m and ε_k^M to the model output uncertainty interval. Defined in this way the output adaptive threshold (cf. Fig. 4.4) should contain the real system response in the fault free mode. An occurrence of the fault is signaled when the system output y_k crosses the output adaptive threshold:

$$\hat{y}_k^m + \varepsilon_k^m \le y_k \le \hat{y}_k^M + \varepsilon_k^M. \tag{4.4}$$

The effectiveness of the suggested method of the robust fault detection requires the determination of a mathematical description of model uncertainty and knowing maximal and minimal values of disturbances ε. For this reason in this chapter the methods of GMDH neural model uncertainty calculation in the form of the model output uncertainty interval are shown. The exact values of the disturbances ε can be calculated following the method described in [152]. However, it should be pointed out that during the application of the

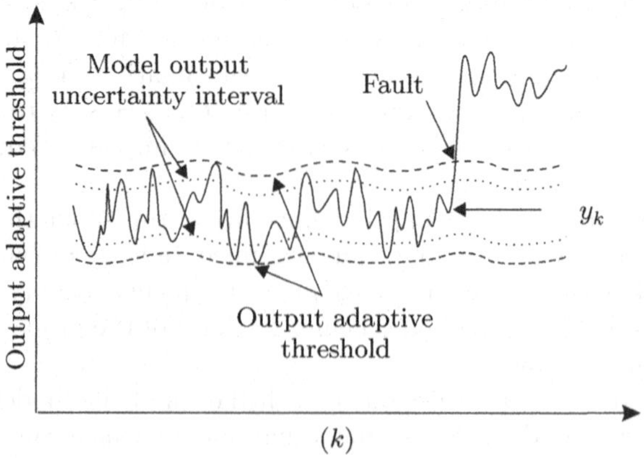

Fig. 4.4. Adaptive threshold in the robust fault detection

fault detection system, the disturbances can not be larger than those which are taken into account in the design of the fault detection system. Not following this rule may cause false alarms in the fault detection. The suggested method of the detection is robust to model uncertainty and disturbances, however, it is crucial to use the representative data sets to a system identification procedure. It prevents an inappropriate working of the fault detection system in case of using new data sets. The problem of the selection of the identification data can be solved by the application of the experiment design techniques [21].

4.3 OBE Algorithm in Parameters Estimation of MLP

The neural network applied during the system identification consists of two layers (cf. Fig. 4.5). The hidden layer includes the neurons with non-linear activation function, however, in the output layer one neuron with linear activation function is employed. Neural model output is then written as follows:

$$\hat{y}_k = \sum_{i=1}^{n_h} \hat{p}_{0,i} f\left(\sum_{j=1}^{n_u} \hat{p}_{i,j} u_{j,k}\right), \tag{4.5}$$

where: $\boldsymbol{u} \in \mathbb{R}^{n_u}$ represents the vector of the model inputs, n_h is the number of the neurons in the hidden layer. It is assumed that the non-linear activation function has the following form $f(\cdot) = \tanh(\cdot)$, i.e., it is a hyperbolic tangent function. It should be strongly underlined that this assumption is not very restrictive and other choices are possible as well. The parameters vector $\hat{\boldsymbol{p}} = [\hat{\boldsymbol{p}}_l^T, \hat{\boldsymbol{p}}_n^T]^T \in \mathbb{R}^{n_h(n_u+1)}$ consists of the parameters of the linear neuron $\hat{\boldsymbol{p}}_l^T = [\hat{p}_{0,1}, \ldots, \hat{p}_{0,n_h}]^T$, and the parameters vectors of the non-linear neurons from the hidden layer $\hat{\boldsymbol{p}}_n^T$, where $\hat{\boldsymbol{p}}_n = [\hat{\boldsymbol{p}}_{1,n}^T, \ldots, \hat{\boldsymbol{p}}_{n_h,n}^T]^T$, $\hat{\boldsymbol{p}}_{i,n} = [\hat{p}_{i,1}, \ldots, \hat{p}_{i,n_u}]^T$. The neural model (4.5) can be described in more condensed form:

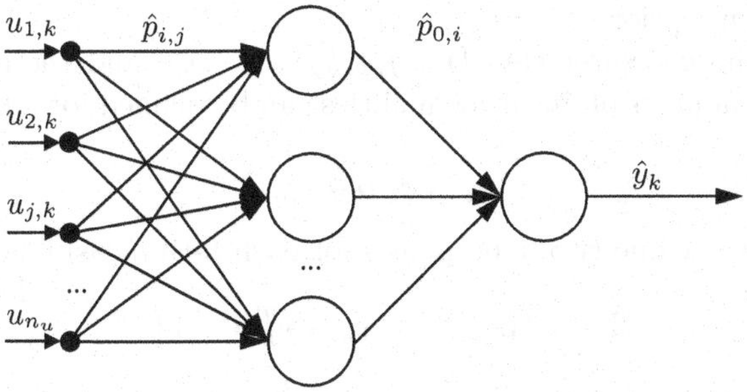

Fig. 4.5. Structure of the MLP for the robust fault detection

$$\hat{y}_k = \boldsymbol{r}_k^T \hat{\boldsymbol{p}}_l, \tag{4.6}$$

where the regressor of the output neuron has the following form:

$$\boldsymbol{r}_k = \left[f(\boldsymbol{u}_k^T \hat{\boldsymbol{p}}_{1,n}), \ldots, f(\boldsymbol{u}_k^T \hat{\boldsymbol{p}}_{n_h,n}) \right]. \tag{4.7}$$

Let us assume that the system output is described as:

$$y_k = \hat{y}_k + \varepsilon_k, \tag{4.8}$$

where y_k is the k–th scalar measurement of the system output and ε_k consists of a structural deterministic error caused by the model-reality mismatch, and the stochastic error caused by the measurement noise.

The problem is to obtain the parameters estimate $\hat{\boldsymbol{p}}$, as well as the associated parameter uncertainty $\mathbb{E}$. This knowledge allows to obtain the uncertainty interval of the model output (4.3). In order to overcome this problem the approaches given in [141] can be applied. Unfortunately, a set of restrictive assumptions concerning the model structure and the linearisation error has to be fulfilled. Additionally, the nature of the disturbances ε_k should be known:

$$\mathcal{E}\left[\boldsymbol{\varepsilon}\right] = \boldsymbol{0}, \tag{4.9}$$

$$\operatorname{cov}\left[\boldsymbol{\varepsilon}\right] = (\sigma)^2 \boldsymbol{I}. \tag{4.10}$$

The assumption (4.9) means that there are no structural errors (deterministic disturbances) and the model uncertainty is described in a purely stochastic way (uncorrelated noise, cf. (4.10)). It must be pointed out that it is rather difficult to satisfy this condition in practice. In order to overcome these problems, in this section a more realistic approach, the BEA presented in Sect. 3.3.2, can be applied. In this algorithm the limitation (4.9) does not exist and there is only the assumption that ε_k is bounded as follows:

$$\varepsilon_k^m \leq \varepsilon_k \leq \varepsilon_k^M, \tag{4.11}$$

where the bounds $\varepsilon_k^m \leq 0$, $\varepsilon_k^M \geq 0$ and $\varepsilon_k^m \neq \varepsilon_k^M$ can be estimated [152], or they are known a'priori [15].

Based on the measurements $\{\boldsymbol{r}_k, y_k\}, k = 1 \ldots n_{\mathcal{T}}$ and the error bounds (4.11) a finite number of linear inequalities can be defined for a linear system in the form:

$$y_k = \boldsymbol{r}_k^T \boldsymbol{p} + \varepsilon_k. \tag{4.12}$$

The inequalities define two hyperplanes for each k-th measurement:

$$\mathbb{H}^+ = \left\{ \mathbf{p} \in \mathbb{R}^{n_p} : y_k - \boldsymbol{r}_k^T \boldsymbol{p} = 1 \right\}, \tag{4.13}$$

and

$$\mathbb{H}^- = \left\{ \mathbf{p} \in \mathbb{R}^{n_p} : y_k - \boldsymbol{r}_k^T \boldsymbol{p} = -1 \right\}. \tag{4.14}$$

Thus, the bounding strip $\mathbb{S}_k$ containing a set of $\boldsymbol{p}$ which satisfies:

$$\mathbb{S}_k = \{\boldsymbol{p} \in \mathbb{R}^{n_p} \mid y_k - \varepsilon_k^M \leq \boldsymbol{r}_k^T \boldsymbol{p} \leq y_k - \varepsilon_k^m\}, \tag{4.15}$$

can be defined. By the intersection of the strips (4.15) for all measurements the feasible parameters set can be obtained. Unfortunately, the polytopic region $\mathbb{E}$ becomes very complicated when the number of measurements and parameters is large, which means that its determination is very complex and time-consuming. Since the number of neurons in the hidden layer n_h of the considered MLP is usually larger, and in the consequence the number of parameters $n_p = n_h(n_u + 1)$ is large as well, the presented approach cannot be directly applied to the parameters estimation of (4.5). A easier solution relies on the approximation of the convex polytope $\mathbb{E}$ by an ellipsoid. For this reason in this section the OBE algorithm described in Sect. 3.3.3 is applied.

4.4 Confidence Estimation of MLP

As contrasted with the linear model in a case of the non-linear MLP the strip $\mathbb{S}$ is defined by a pair of hiperspaces. Figure 4.6 presents two pairs of hiperspaces defining the set of $\mathbb{E}$. The intersection of hyperspaces results in a feasible parameters set. Contrary to the linear model, in the case of the MLP the region $\mathbb{E}$ can not be convex or even disconnected which causes that the model is not uniquely identifiable. Thus, it should be strongly underlined that the obtained ellipsoid constitutes a local approximation of the real parameter uncertainty.

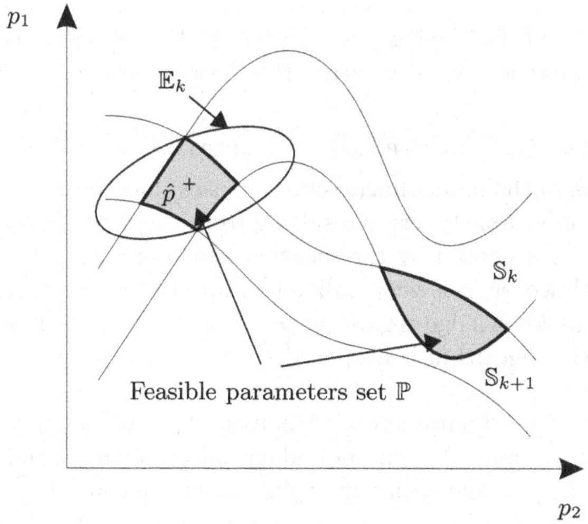

Fig. 4.6. Hypothetical feasible parameters set for a non-linear model

In order to employ the proposed OBE algorithm to the parameters estimation of the MLP it is necessary to linearize the model (4.5) around the parameters estimate $\hat{\boldsymbol{p}}$.

$$\begin{aligned}\hat{y}_{l,k} &= \hat{y}_k + \nabla\hat{y}|_{\boldsymbol{p}=\hat{\boldsymbol{p}}_{k-1}}(\boldsymbol{p}-\hat{\boldsymbol{p}}_k)\\ &= \hat{\boldsymbol{r}}_k^T\hat{\boldsymbol{p}}_{l,k} + \nabla\hat{y}_l|_{\boldsymbol{p}=\hat{\boldsymbol{p}}_{k-1}}(\boldsymbol{p}_l-\hat{\boldsymbol{p}}_{l,k})\\ &+ \nabla\hat{y}_n|_{\boldsymbol{p}=\hat{\boldsymbol{p}}_{k-1}}(\boldsymbol{p}_n-\hat{\boldsymbol{p}}_{n,k})\\ &= \hat{\boldsymbol{r}}_k^T\boldsymbol{p}_l + \nabla\hat{y}_n|_{\boldsymbol{p}=\hat{\boldsymbol{p}}_{k-1}}\boldsymbol{p}_n - \nabla\hat{y}_n|_{\boldsymbol{p}=\hat{\boldsymbol{p}}_{k-1}}\hat{\boldsymbol{p}}_{n,k}.\end{aligned} \tag{4.16}$$

If it is assumed, that $\boldsymbol{p}=[\boldsymbol{p}_l^T,\boldsymbol{p}_n^T]^T$, and $\nabla\hat{y}=\left[\hat{\boldsymbol{r}}_k^T,\nabla\hat{y}_n\right]$, where $\nabla\hat{y}$ can be obtained as follows:

$$\begin{aligned}\nabla\hat{y} = \Big[&\hat{\boldsymbol{r}}_k^T, p_{0,1}u_1\left(1-f(\boldsymbol{u}^T\boldsymbol{p}_{1,n})^2\right),\ldots,\\ &p_{0,1}u_{n_u}\left(1-f(\boldsymbol{u}^T\boldsymbol{p}_{1,n})^2\right),\ldots,\\ &p_{0,n_h}u_1\left(1-f(\boldsymbol{u}^T\boldsymbol{p}_{n_h,n})^2\right),\ldots,\\ &p_{0,n_h}u_{n_u}\left(1-f(\boldsymbol{u}^T\boldsymbol{p}_{n_h,n})^2\right)\Big],\end{aligned} \tag{4.17}$$

then (4.16) receives the following form:

$$\hat{y}_{l,k} = \nabla\hat{y}|_{\boldsymbol{p}=\hat{\boldsymbol{p}}_{k-1}}\boldsymbol{p} + \nabla\hat{y}_n|_{\boldsymbol{p}=\hat{\boldsymbol{p}}_{k-1}}\hat{\boldsymbol{p}}_{n,k} = \boldsymbol{r}_k^T\boldsymbol{p} + \nabla\hat{y}_n|_{\boldsymbol{p}=\hat{\boldsymbol{p}}_{k-1}}\hat{\boldsymbol{p}}_{n,k}. \tag{4.18}$$

The relation between the non-linear model (4.5), and its linear equivalent (4.18), can be described as:

$$\hat{y}_k = \hat{y}_{l,k} + o(\boldsymbol{p},\hat{\boldsymbol{p}}), \tag{4.19}$$

where $o(\boldsymbol{p},\hat{\boldsymbol{p}})$ stands for the higher-order terms of the Taylor series expansion. In order to apply the OBE algorithm the system output should be written as:

$$y_k = \hat{y}_k^l + \varepsilon_k = \nabla\hat{y}|_{\boldsymbol{p}=\hat{\boldsymbol{p}}_{k-1}}\boldsymbol{p} - \nabla\hat{y}_n|_{\boldsymbol{p}=\hat{\boldsymbol{p}}_{k-1}}\hat{\boldsymbol{p}}_{n,k-1} + \varepsilon_k. \tag{4.20}$$

The application of the model linearisation causes that the error ε_k consists of the structural deterministic error resulting from the model-reality mismatch, the stochastic error caused by the measurement noise and also linearisation error $o(\boldsymbol{p},\hat{\boldsymbol{p}})$. However, it is very difficult to obtain the exact value of $o(\boldsymbol{p},\hat{\boldsymbol{p}})$ as it requires the knowledge regarding $\boldsymbol{p}$. To overcome this difficulty, the following sequential algorithm is proposed:

Input : $\hat{\boldsymbol{p}}_0$ – the initial parameters estimates, $\boldsymbol{P}_0$ – the matrix describing the initial ellipsoid, ε_0^m and ε_0^M - the boundary values of the initial output error
Output : $\varepsilon_{n_T}^m$ i $\varepsilon_{n_T}^M$ – the boundary values of the output error interval

1. Calculate $\hat{\boldsymbol{p}}$ and $\boldsymbol{P}$ with the application of the OBE algorithm.
2. If $\det(\boldsymbol{P}_0) \leqslant \det(\boldsymbol{P})$ then set $\hat{\boldsymbol{p}} = \hat{\boldsymbol{p}}_0$, $\boldsymbol{P} = \boldsymbol{P}_0$ and STOP, else go to step 3.
3. Obtain:

$$\varepsilon_k^m = \min_{k=1,\dots n_T} y_k - \hat{y}_k,$$

$$\varepsilon_k^M = \max_{k=1,\dots n_T} y_k - \hat{y}_k.$$

4. Set $\hat{\boldsymbol{p}}_0 = \hat{\boldsymbol{p}}$, $\boldsymbol{P}_0 = \boldsymbol{P}$ and go to step 1.

The above presented algorithm works until the reduction of the model uncertainty succeeds. The volume of the model uncertainty is defined by the determinant of the ellipsoid $\boldsymbol{P}$. Knowing the parameters estimate $\hat{\boldsymbol{p}}$ as well as the matrix $\boldsymbol{P}$ it is possible to obtain the bounds of the linear model output uncertainty interval:

$$\hat{y}_{l,k}^m = \nabla \hat{y}^T \hat{\boldsymbol{p}}_l - \sqrt{\nabla \hat{y} \boldsymbol{P} \nabla \hat{y}^T}, \tag{4.21}$$

$$\hat{y}_{l,k}^M = \nabla \hat{y}^T \hat{\boldsymbol{p}}_l + \sqrt{\nabla \hat{y} \boldsymbol{P} \nabla \hat{y}^T}. \tag{4.22}$$

In order to get the output uncertainty interval of the whole model, it is necessary to calculate the bounds of $o(\boldsymbol{p}, \hat{\boldsymbol{p}})$ based on the expression (4.19):

$$\begin{aligned} o(\boldsymbol{p}, \hat{\boldsymbol{p}}) &= y_k - \hat{y}_{l,k} = \boldsymbol{r}_{0,k}\boldsymbol{p}_l - \hat{\boldsymbol{r}}_k \boldsymbol{p}_l - \nabla \hat{y}_n|_{\boldsymbol{p}=\hat{\boldsymbol{p}}_{k-1}} (\boldsymbol{p}_n - \hat{\boldsymbol{p}}_{n,k}) \\ &= (\boldsymbol{r}_{0,k} - \hat{\boldsymbol{r}}_k)\boldsymbol{p}_l - \nabla \hat{y}_n|_{\boldsymbol{p}=\hat{\boldsymbol{p}}_{k-1}} (\boldsymbol{p}_n - \hat{\boldsymbol{p}}_{n,k}). \end{aligned} \tag{4.23}$$

Assuming that the parameters errors can be defined as:

$$\boldsymbol{e}_{l,k} = \boldsymbol{p}_l - \hat{\boldsymbol{p}}_{l,k}, \tag{4.24}$$

and

$$\boldsymbol{e}_{n,k} = \boldsymbol{p}_n - \hat{\boldsymbol{p}}_{n,k}. \tag{4.25}$$

where: $\boldsymbol{e}_{n,k} = [\boldsymbol{e}_{1,n,k}, \dots, \boldsymbol{e}_{n_h,n,k}]$, $\boldsymbol{e}_{i,n} = [e_{i,1,k}, \dots, e_{i,n_u,k}]$, $\boldsymbol{e}_k = \boldsymbol{p} - \hat{\boldsymbol{p}}_k$, the linearisation error as the result of the substitution (4.24) to (4.23) can be written as:

$$\begin{aligned} o(\boldsymbol{p}, \hat{\boldsymbol{p}}) &= (\boldsymbol{r}_{0,k} - \hat{\boldsymbol{r}}_k)(\boldsymbol{e}_{l,k} + \hat{\boldsymbol{p}}_{l,k}) - \nabla \hat{y}_n|_{\boldsymbol{p}=\hat{\boldsymbol{p}}_{k-1}} (\boldsymbol{p}_n - \hat{\boldsymbol{p}}_{n,k}) \\ &= \Delta \boldsymbol{r}(\boldsymbol{e}_{l,k} + \hat{\boldsymbol{p}}_{l,k}) - \nabla \hat{y}_n|_{\boldsymbol{p}=\hat{\boldsymbol{p}}_{k-1}} (\boldsymbol{p}_n - \hat{\boldsymbol{p}}_{n,k}) \\ &= \Delta \boldsymbol{r}(\boldsymbol{e}_{l,k} + \hat{\boldsymbol{p}}_{l,k}) - \nabla_n(\boldsymbol{e}_{n,k}). \end{aligned} \tag{4.26}$$

The expression $\Delta \boldsymbol{r} = (\boldsymbol{r}_{0,k} - \hat{\boldsymbol{r}}_k)$ in (4.26) can be obtained as follows:

$$\begin{aligned}\Delta\boldsymbol{r} = & \left[f\left(\sum_{j=1}^{n_u} p_{1,j}u_{j,k}\right) - f\left(\sum_{j=1}^{n_u} \hat{p}_{1,j,k}u_{j,k}\right), \ldots, \right. \\ & \left. f\left(\sum_{j=1}^{n_u} p_{n_h,j}u_{j,k}\right) - f\left(\sum_{j=1}^{n_u} \hat{p}_{n_h,j,k}u_{j,k}\right) \right] = \\ & \left[f\left(\sum_{j=1}^{n_u} \left(e_{1,j,k} - \hat{p}_{1,j,k}\right) u_{j,k}\right) - f\left(\sum_{j=1}^{n_u} \hat{p}_{1,j,k}u_{j,k}\right), \ldots, \right. \\ & \left. f\left(\sum_{j=1}^{n_u} \left(e_{n_h,j,k} - \hat{p}_{n_h,j,k}\right) u_{j,k}\right) - f\left(\sum_{j=1}^{n_u} \hat{p}_{n_h,j,k}u_{j,k}\right) \right].\end{aligned} \tag{4.27}$$

Substituting (4.27) to (4.26):

$$\begin{aligned}o(\boldsymbol{p}, \hat{\boldsymbol{p}}) = & \sum_{i=1}^{n_h} (e_{l,i,k} + \hat{p}_{0,i,k}) \cdot \\ & \cdot \left(f\left(\sum_{j=1}^{n_u} \left(e_{i,j,k} - \hat{p}_{i,j,k}\right) u_{j,k}\right) - f\left(\sum_{j=1}^{n_u} \hat{p}_{i,j,k}u_{j,k}\right)\right) - \\ & - \sum_{i=1}^{n_h} \hat{p}_{0,i,k} \left(1 - f\left(\sum_{j=1}^{n_u} \hat{p}_{i,j,k}u_{j,k}\right)^2\right) \cdot \sum_{j=1}^{n_u} e_{i,j,k}u_{j,k}\end{aligned} \tag{4.28}$$

and assuming:

$$\alpha_i = \sum_{j=1}^{n_u} e_{i,j,k}u_{j,k}, \tag{4.29}$$

$$\beta_i = \sum_{j=1}^{n_u} \hat{p}_{i,j,k}u_{j,k}, \tag{4.30}$$

then (4.28) takes the following form:

$$\begin{aligned}o(\boldsymbol{p}, \hat{\boldsymbol{p}}) = & \sum_{i=1}^{n_h} (e_{l,i,k} + \hat{p}_{0,i,k}) \left(f\left(\alpha_i - \beta_i\right) - f\left(\beta_i\right)\right) \\ & - \sum_{i=1}^{n_h} \hat{p}_{0,i,k} \left(1 - f\left(\beta_i\right)^2\right) \alpha_i = \\ & = \sum_{i=1}^{n_h} \left[(e_{l,i,k} + \hat{p}_{0,i,k}) \left(f\left(\alpha_i - \beta_i\right) - f\left(\beta_i\right)\right) - \right. \\ & \left. - \hat{p}_{0,i,k} \left(1 - f\left(\beta_i\right)^2\right) \alpha_i \right]\end{aligned} \tag{4.31}$$

Moreover, it is assumed that:

$$1 - f(\beta_i)^2 = \frac{1}{\cosh(\beta_i)^2}, \tag{4.32}$$

and

$$f(\alpha_i - \beta_i) - f(\beta_i) = \frac{\sinh(\alpha_i)}{\cosh(\alpha_i + \beta_i)\cosh(\beta_i)}, \tag{4.33}$$

then substituting expressions (4.32) and (4.33) to (4.31) the final form of the linearisation error can be written as:

$$o(\boldsymbol{p}, \hat{\boldsymbol{p}}) = \sum_{i=1}^{n_h} \left[(e_{l,i,k} + \hat{p}_{0,i,k}) \frac{\sinh(\alpha_i)}{\cosh(\alpha_i + \beta_i)\cosh(\beta_i)} - \frac{\hat{p}_{0,i,k}\alpha_i}{\cosh(\beta_i)^2} \right]. \tag{4.34}$$

Without a loss of generalization it is assumed that $u \geq 0$ for $i = 1, \ldots, n_{\mathcal{T}}$. On the basis of the expression (4.34) it can be seen that the linearisation error $o(\boldsymbol{p}, \hat{\boldsymbol{p}})$ depends on the value $e_{i,k}$. This value is overbounded by the square roots of the diagonal elements matrix $\boldsymbol{P}$ which defines the size and the orientation of the ellipsoid:

$$-\sqrt{P_{i,i}} \leq e_{i,k} \leq \sqrt{P_{i,i}}, \tag{4.35}$$

where: $i = 1, \ldots, n_h(1 + n_u)$. Thus, depending on the changes of the values $e_{i,k}$, also the value $o(\boldsymbol{p}, \hat{\boldsymbol{p}})$ is changed in the limited interval:

$$o(\boldsymbol{p}, \hat{\boldsymbol{p}})^m \leq o(\boldsymbol{p}, \hat{\boldsymbol{p}}) \leq o(\boldsymbol{p}, \hat{\boldsymbol{p}})^M. \tag{4.36}$$

Based on the expressions (4.3), (4.21), (4.22) and (4.36) it is possible to obtain the neural model output uncertainty interval:

$$\begin{aligned} &\hat{\boldsymbol{r}}_k^T \hat{\boldsymbol{p}}_l - \sqrt{\nabla \hat{y} \boldsymbol{P} \nabla \hat{y}^T} + o(\boldsymbol{p}, \hat{\boldsymbol{p}})^m \leq \hat{y}_k \leq \\ &\hat{\boldsymbol{r}}_k^T \hat{\boldsymbol{p}}_l + \sqrt{\nabla \hat{y} \boldsymbol{P} \nabla \hat{y}^T} + o(\boldsymbol{p}, \hat{\boldsymbol{p}})^M, \end{aligned} \tag{4.37}$$

which is presented in Fig. 4.7.

The model output uncertainty interval, calculated with the application of the MLP model, should contain the real system response in a fault free mode. The range of the model output confidence interval depends on the MLP uncertainty. As the measurements y_k of the diagnosed system are corrupted by the noise and it is not possible to obtain their exact value, it is necessary to modify the model output uncertainty interval (4.37). This modification relies on adding the boundary values of the output error (4.11) to the model output uncertainty interval (4.37):

$$\begin{aligned} &\hat{\boldsymbol{r}}_k^T \hat{\boldsymbol{p}}_l - \sqrt{\nabla \hat{y} \boldsymbol{P} \nabla \hat{y}^T} + o(\boldsymbol{p}, \hat{\boldsymbol{p}})^m + \varepsilon_{n_{\mathcal{T}}}^m \leq y_k \leq \\ &\hat{\boldsymbol{r}}_k^T \hat{\boldsymbol{p}}_l + \sqrt{\nabla \hat{y} \boldsymbol{P} \nabla \hat{y}^T} + o(\boldsymbol{p}, \hat{\boldsymbol{p}})^M + \varepsilon_{n_{\mathcal{T}}}^M. \end{aligned} \tag{4.38}$$

The newly defined output adaptive threshold is presented in Fig. 4.8. The occurrence of the fault is signaled when the system output signal crosses the output adaptive threshold.

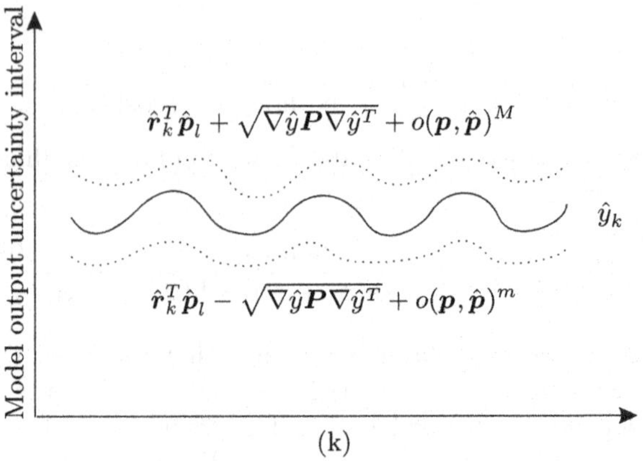

Fig. 4.7. MLP output uncertainty interval

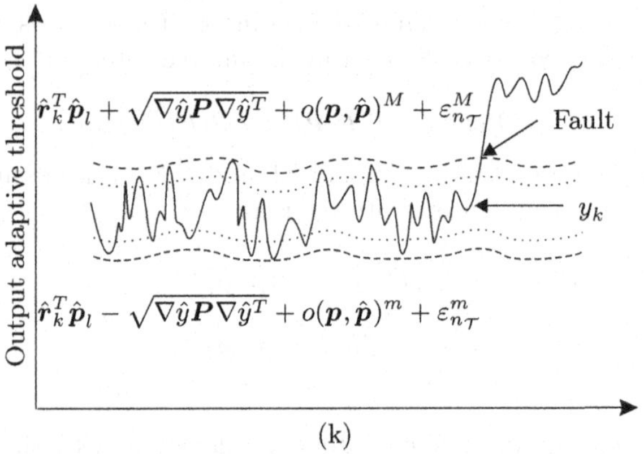

Fig. 4.8. MLP output adaptive threshold

4.5 An Illustrative Example - NLMS vs. OBE

The aim of this section is to show the effectiveness of the proposed approach on the basis of the OBE algorithm in the task of the parameters estimation of the static non-linear system. Let us consider the following system:

$$y_k = p_1 \sin(p_2 u_k) + \varepsilon_k, \tag{4.39}$$

where the nominal values of parameters are $\boldsymbol{p}_n = [1.2, 0.5]^T$, the input data u_k, $k = 1 \ldots n_{\mathcal{T}}$ and the noise ε_k, $k = 1 \ldots n_{\mathcal{T}}$ are generated according to the uniform distribution, i.e., $u_k \in \mathcal{U}(0.1, 1.2)$ and $\varepsilon_k \in \mathcal{U}(-0.05, 0.1)$. The problem is to obtain the parameters estimate $\hat{\boldsymbol{p}}$ and the corresponding feasible parameters set using the set of input-output measurements $\{u_k, y_k\}_{k=1}^{n_{\mathcal{T}}=1000}$ and the OBE algorithm. It is assumed that initial conditions for the algorithm are: $\hat{\boldsymbol{p}}_0 = [0.1, 0.1]^T$, $\boldsymbol{P}_0 = 200\boldsymbol{I}_{n_p}$, $\varepsilon_0^m = -0.6$, $\varepsilon_0^M = 0.6$. For the purpose of a comparison, a Matlab function called `lsqcurvefit` is employed. This function implements the NLMS method. The results of the comparison are shown in Tab. 4.1.

Table 4.1. Results of a comparative study - the OBE vs. NLMS

	$\hat{p}$	$\|\boldsymbol{e}\|_1$	$\|\boldsymbol{e}\|_2$
OBE	[1.2245, 0.4997]	0.0248	0.0245
NLMS	[1.2224, 0.5135]	0.0359	0.0261

From these results it can be seen that the proposed approach provides good results. As it was expected, the proposed sequential algorithm makes it possible to decrease the volume of the ellipsoid which describes the model uncertainty (cf. Figs. 4.9 and 4.10). It should also be mentioned that the final values of $\varepsilon_{n_{\mathcal{T}}}^m$ and $\varepsilon_{n_{\mathcal{T}}}^M$ are -0.063 and 0.101, respectively (note the similarity to $\mathcal{U}(-0.05, 0.1)$).

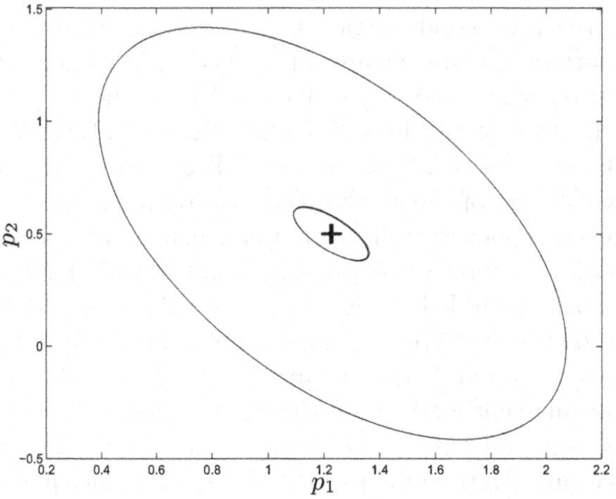

Fig. 4.9. Consecutive ellipses obtained with the sequential algorithm

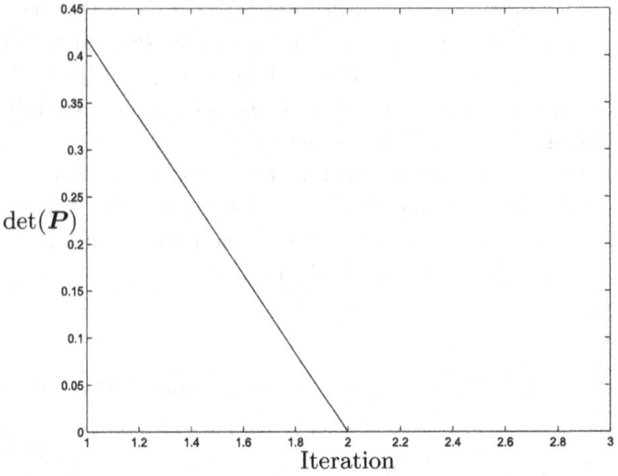

Fig. 4.10. Determinant of matrix describing the feasible parameters set

4.6 MLP in Robust Fault Detection of Valve Actuator

In order to show the effectiveness of the robust fault detection system based on the MLP model the Research Training Network on Development and Application of Methods for Actuator Diagnosis in Industrial Control Systems (RTN DAMADICS) benchmark is employed [156, 157]. The RTN DAMADICS is focused on the diagnosis of a 5-stage evaporation and boiler station process in the Lublin Sugar Factory S.A. in Poland.

The evaporation station consists of five serially connected evaporators in which the required juice condensation is achieved by extracting water from the juice. The obtained steam is transported to the subsequent sections of the evaporation station where the steam delivered from the local power station is used to the heating of the juice from the sugar beat. Thus, the juice is gradually thickened as a result of the controlled juice boiling process. The thick juice down stream of the evaporation station is pumped into the set of heated tanks where sugar crystallisation takes place. Sugar crystallisation is a batch processes. To engage this process, a portion of artificially prepared crystallisation nuclei is added to the thick juice filling up the heated tanks. The crystallisation process typically takes some hours. As a result, a bi-phase mixture of sugar crystals and liquid caramel is developed. When the crystallisation process is sufficiently advanced the bi-phase mixture from a particular tank is pumped out and fed into the intermediate tank. The tank is ready again to be filled up with the next portion of the thick juice and thereby start the next sugar crystallisation process. The content of the intermediate tank is periodically fed into centrifuges separating sugar crystals from caramel.

After that, the sugar crystals are graded by means of vibrating screens and directed for the final stages of confectionary production [156].

The process of the juice evaporation is controlled by acting on flows of the thin and thick juice at the input and output of the evaporation station. This process is highly disturbed by the fluctuation of thin juice density and uncontrolled power demands. Furthermore, it is significant as several hazards can be developed: juice can overheated, sugar can be lost, caramel sedimentation can disturb the efficiency of the process, allowable vapour pressures can be exceeded. Indeed the process can shut-down due to faulty behaviour of actuators controlling the juice flow through the installation [156]. Among numerous control systems in the evaporation station, the control of temperatures, steam flow, pressures of the juice and juice flows through the installation can be distinguished.

For the fault detection purpose the three benchmark actuators can be chosen: the first controlling thin juice inflow (cf. Fig. 4.11), and the second controlling thick juice outflow (cf. Fig. 4.12) located in the evaporation station. The third benchmark actuator is placed in the power station boiler water supply system (cf. Fig. 4.13).

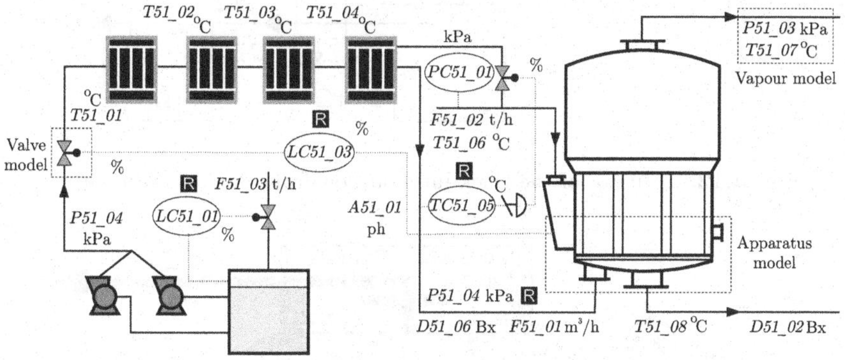

Fig. 4.11. Actuator of the thin juice inflow into the first evaporator

In this section for the validation of the developed fault detection method, the actuator $FC57\_03$ (cf. Fig. 4.14) controlling thick juice outflow from the last evaporation station is chosen. The considered electro-pneumatic valve actuator is widespread in industrial environment. These devices are vulnerable in the unfavorable environmental conditions. Unfortunately, the influence on high temperatures, humidity, pollution, chemical solvents and aggressive media may result in the numerous faults. The early detection of small or incipient faults of the actuators and their replacement may prevent serious damages of the evaporation station and break-off of the industrial process.

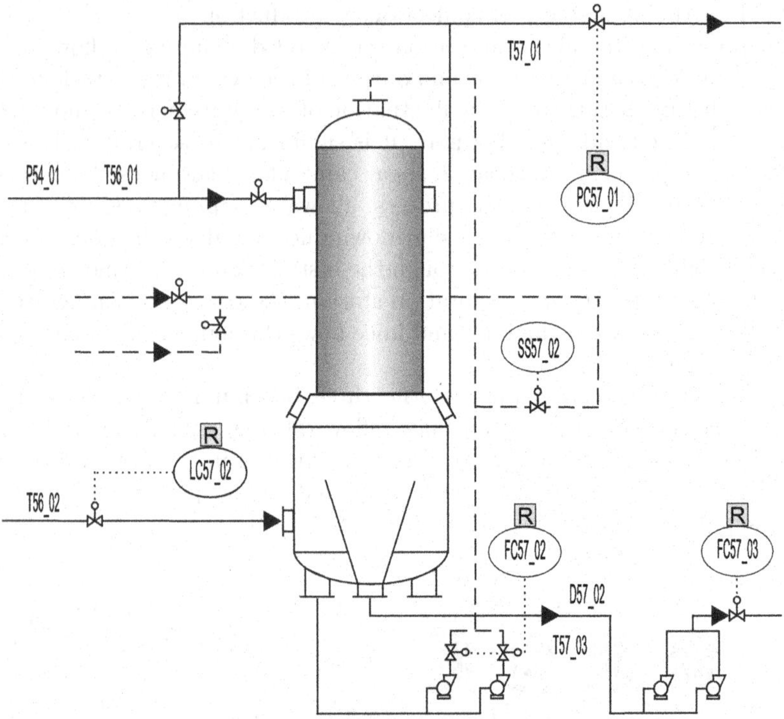

Fig. 4.12. Actuator of the thick juice outflow from the last evaporator

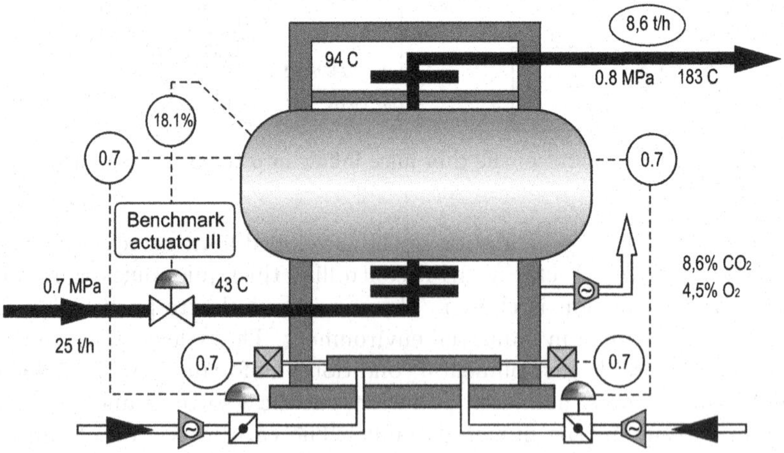

Fig. 4.13. Actuator of the water inflow into the steam generator boiler

Fig. 4.14. An electro-pneumatic valve actuator

The element selected for modelling and fault detection is a final control element which interacts with the controlled process. The scheme of the diagnosed actuator with all the available process variables is shown in Fig. 4.15, where V_1, V_2 and V_3 denote the bypass valves, ACQ and CPU are the data

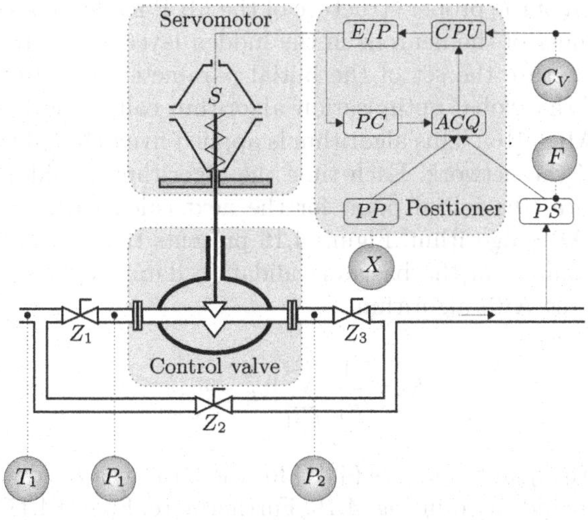

Fig. 4.15. Diagram of an electro-pneumatic valve actuator

acquisition and positioner central processing units. E/P and FT are the electro-pneumatic and value flow rate transducers. Finally, DT and PT represent the displacement and pressure transducers. On the ground of the process analysis the following model of the juice flow at the outlet of the valve is chosen for the modelling:

$$F = r_F(X, P_1, P_2, T_1), \tag{4.40}$$

where $r_F(\cdot)$ denotes the modelled relationship, X is the rod displacement, P_1 and P_2 are the pressures at the inlet and outlet of the valve respectively, and T_1 represents the juice temperature at the inlet of the valve. A detailed description regarding the data and the artificially introduced faults can be found in Tab. 4.2. Unfortunately, the data turn out to be sampled too fast. Thus, every 10-th value is picked resulting in the 1000-th elements training and selection data sets.

Table 4.2. List of training, validation and faults data sets from the valve actuator

Fault	Range (samples)	Fault/data description
No fault	1–10000	Training data set
No fault	10001–20000	Validation data set
f_{16}	57475–57530	Positioner supply pressure drop
f_{17}	53780–53794	Unexpected pressure drop across the valve
f_{18}	54600–54700	Partly opened bypass valves
f_{19}	55977–56015	Flow rate sensor fault

The selection of the proper structure of the MLP model relies on gradually increasing number of the neurons in the hidden layer from 1 to 20. For each network architecture the set of the initial parameters is obtained with the application of the global optimisation algorithm called the Adaptive Random Search (ARS) [16]. This algorithm is applied five hundred times for each architecture of the network. Each time the algorithm provides the solution which constitutes the initial point for the next calculation with the application of the ARS algorithm. Figure 4.16 presents the values of the quality index $B_\mathcal{V}$ calculated on the basis of validation data set for each successive application of the ARS algorithm.

$$B_\mathcal{V} = \frac{1}{n_\mathcal{V}} \sum_{k=1}^{n_\mathcal{V}} |y_k - \hat{y}_k|, \tag{4.41}$$

The values of the quality indexes (4.41) for the best, mean and worst network architecture are indicated in Fig. 4.16. Furthermore, Figure 4.17 presents the best value of the quality index $B_\mathcal{V}$ for all 20 architectures of the MLP neural network.

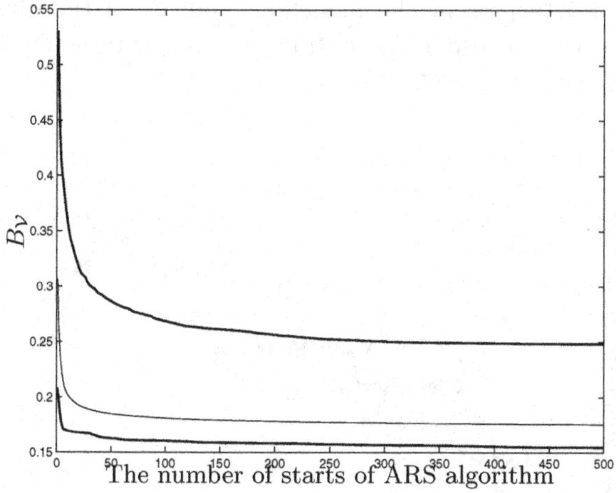

Fig. 4.16. $B_\mathcal{V}$ obtained during successive using of the ARS algorithm

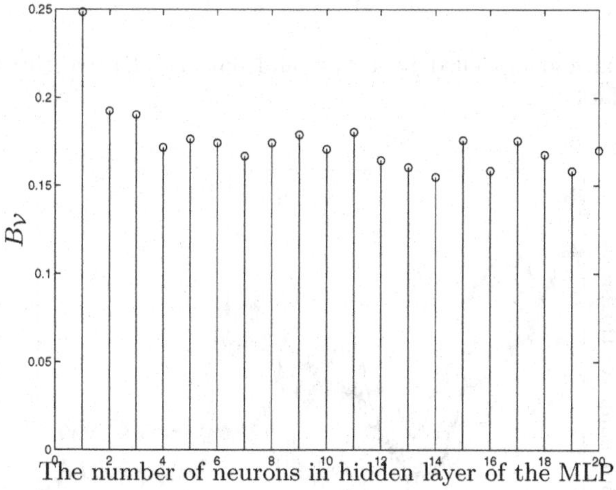

Fig. 4.17. Relation between $B_\mathcal{V}$ and the number of neurons in the hidden layer of the MLP

This procedure is necessary because as it is mentioned in Sect. 4.4, the application of the OBE algorithm to the parameters estimation requires the lineariston of the model (4.5) around a parameters estimate $\hat{\boldsymbol{p}}$. Thus, a relatively, good initial estimate is required. Based on the obtained results, the neural network with $n_h = 14$ neurons in the hidden layer, which provides the best modelling quality, is chosen to the robust fault detection. In the next step, the algorithm described in Sect. 4.4 is employed and the following bounds of the output error are obtained: $\varepsilon_{n_\mathcal{T}}^m = -0.8298$ and $\varepsilon_{n_\mathcal{T}}^M = 0.6371$.

Figure 4.18 and 4.19 present the modelling abilities of the obtained MLP model as well as the corresponding output adaptive threshold obtained with the approach proposed in Sect. 4.4.

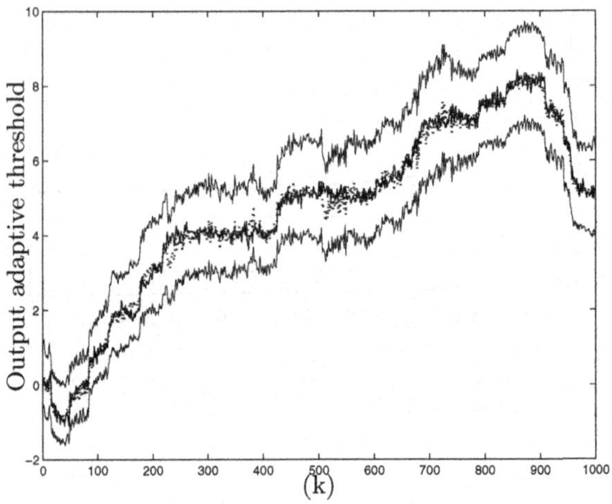

Fig. 4.18. Real and modelled juice flow and output adaptive threshold for the identification data

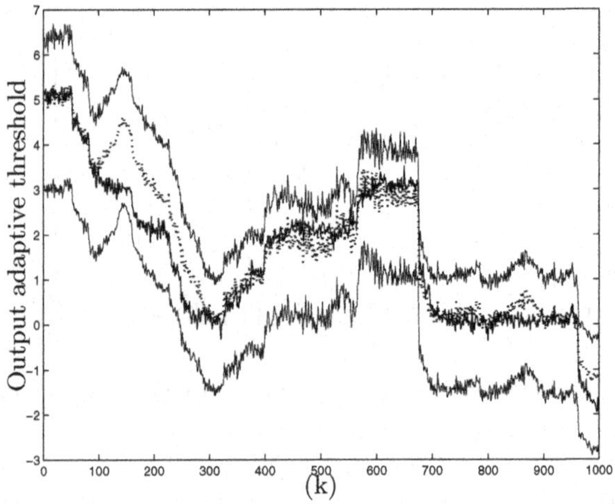

Fig. 4.19. Real and modelled juice flow and output adaptive threshold for the validation data

Knowing the model structure and possessing the knowledge regarding its uncertainty it is possible to design a robust fault detection system. In order to achieve this goal, the output adaptive threshold has to be calculated according to (4.38) defined in Sect. 4.4. For the fault detection purpose the fault scenarios $f_{16} - f_{19}$ defined in Tab. 4.2 are applied. Figures 4.20-4.23 present the real system response (juice flow) as well as the corresponding output adaptive threshold obtained with the MLP model for scenarios $f_{16} - f_{19}$.

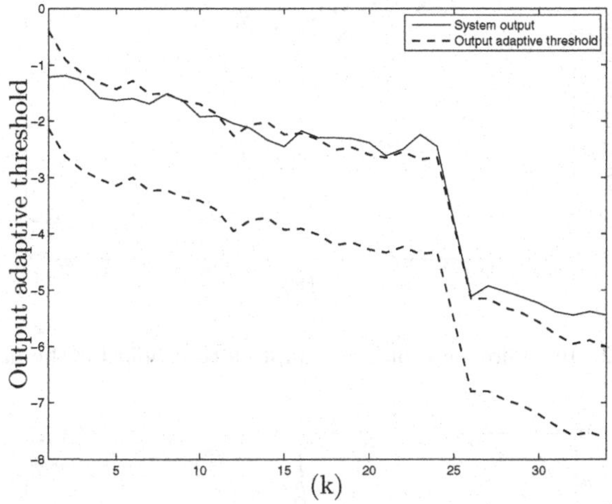

Fig. 4.20. Juice flow and output adaptive threshold for the fault f_{16}

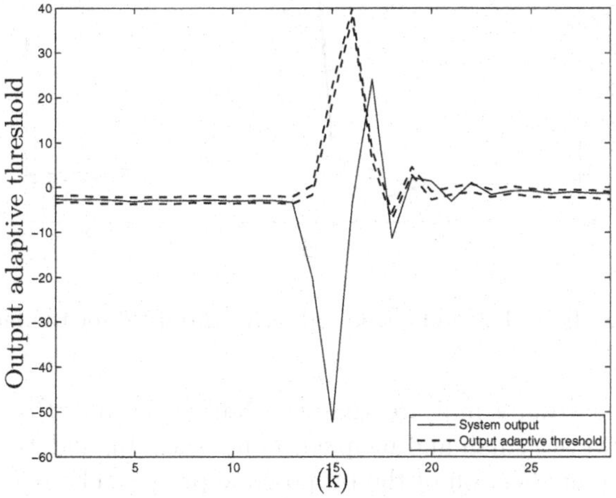

Fig. 4.21. Juice flow and output adaptive threshold for the fault f_{17}

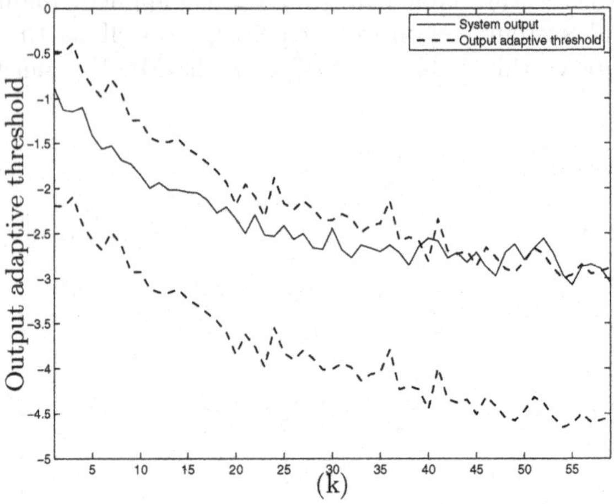

Fig. 4.22. Juice flow and output adaptive threshold for the fault f_{18}

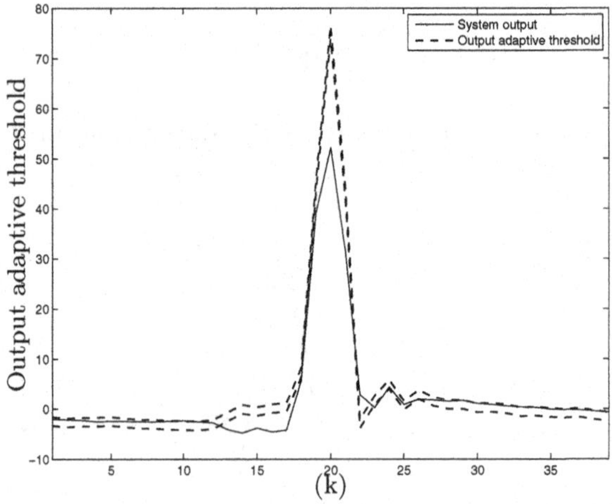

Fig. 4.23. Juice flow and output adaptive threshold for the fault f_{19}

The obtained results indicate that the faults f_{17} and f_{19} are detected very easily. However, small and incipient faults such f_{16} and f_{18} are difficult to detect by the application of the proposed approach. The reason of such a situation is that the system output bounds obtained with the OBE algorithm are too large and hence the sensitivity to faults is not good enough. This means that in the future research direction it is necessary to employ a more accurate technique than the OBE algorithm.

For the sake of comparison the traditional residual-based fault detection approach is applied. As a result the residual signal obtained with the application of the MLP network is presented in Fig. 4.24.

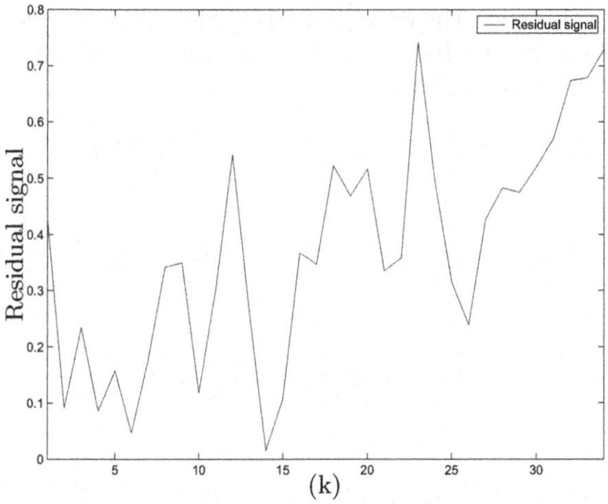

Fig. 4.24. Residuum obtained with the MLP for the fault f_{16}

From the comparison of Figs. 4.20 and 4.24 it can be noticed that the proposed method is superior to traditional residual-based fault detection approach. It follows from the fact that residual-based fault detection scheme makes it impossible to detect the fault f_{16}.

4.7 Concluding Remarks

The aim of this chapter is concerned with obtaining neural models and calculating their uncertainty directly from the observed data. It is shown how to estimate the parameters of the MLP neural network and the corresponding uncertainty. Based on the designed neural model a novel robust fault detection scheme is proposed which may support the diagnostic decisions. Another objective is to provide an industrial application study of the proposed fault detection scheme for a chosen part of the evaporation station in the Lublin sugar factory in Poland. It is shown, using a set of faults, the proposed MLP model-based fault detection scheme can provide good results and all considered faults can be detected. However, it should be pointed out that the proposed approach has the same drawback as the OBE algorithm, i.e., the feasible parameters set obtained with this approach is only an approximation of the original one and may be not convex or even partly separated. It causes that the model is not uniquely identifiable. The approximation of the original feasible parameters set by the ellipsoid leads to the determination of too wide output adaptive threshold which is used to the robust fault detection. For this reason some small faults can be difficult to detect. Thus, more accurate estimation technique than the OBE algorithm should be applied. For that purpose the so-called zonotopic representation can be used instead of the ellipsoidal one. In this case, the exact bounds of the parameter uncertainty can be obtained. However, a large increase in the memory and time expenditure are the cost of this modification.

5

GMDH Networks in Robust Fault Detection of Dynamic Non-linear Systems

5.1 Introduction

In the case of the classical ANNs such as the MLP, the problem reduces to the selection of the number of layers and the neurons in a particular layer. If the obtained network does not satisfy prespecified requirements, then a new network structure is selected and the parameters estimation is repeated once again. The determination of the appropriate structure and parameters of the model in the presented way is a complex task. Furthermore, an arbitrary selection of the ANN structure can be a source of the model uncertainty. Thus, it seems desirable to have a tool, which can be employed to the automatic selection of the ANN structure, based only on the measured data. To overcome this problem the GMDH neural networks [69, 70] have been proposed. The synthesis process of the GMDH neural network is based on the iterative processing of a sequence of operations. This process leads to the evolution of the resulting model structure in such a way so as to obtain the best quality approximation of the identified system. Thus, the task of designing a neural network is defined in order to obtain a model with small uncertainty.

Apart from the contribution to the model uncertainty of the errors caused by an inappropriate structure selection also parameters estimates inaccuracy influences on the model quality. This problem is widely presented in the paper [152], where the potential sources of the model uncertainty following from the application of an improper parameters estimation algorithm are described. In particular, it is shown that usual parameters estimation algorithms work on the basis of the incorrect assumptions concerning the properties of the noises and disturbances which affect on the data used during system identification. In order to overcome this problem the BEA [15, 16] presented in Sect. 3.3.2 can be applied. This approach is based on a more realistic assumption that the noises and disturbances lie between given prior bounds. Unfortunately, in spite of many advantages this algorithm has also significant memory and time expenditure. It follows from the fact that the BEA estimates the model

M. Mrugalski, *Advanced Neural Network-Based Computational Schemes for Robust Fault Diagnosis*, Studies in Computational Intelligence 510,
DOI: 10.1007/978-3-319-01547-7_5,

parameters and their uncertainty in the form of an admissible parameter set, which shape and size depend on the number of parameters and identification data. In the case of a neural model, which consists of a great number of neurons, the determination of parameters estimations is not possible considering significant memory and time expenditure. In order to overcome this disadvantage the OBE or ZBA algorithms can be used. The application of these algorithms allows to estimate the parameters of the GMDH neural network and the model uncertainty in a form of the ellipsoid or zonotope which are an approximation of the exact admissible parameter set. Based on this result it is possible to define the model uncertainty in the form of the output adaptive thresholds, which can be applied in the numerous industrial applications, e.g., robust fault detection.

The chapter is organized as follows: Section 5.2 contains original developments regarding the uncertainty estimation of the GMDH neural network. In particular, the methods of neural model uncertainty with the LMS, BEA, OBE and ZBA algorithms are developed. Section 5.3 presents the modification of the original GMDH algorithm leading to decrease of the GMDH model uncertainty. In Sect. 5.4 the details of additional improvement of the GMDH neural model are presented. Section 5.5 contains an illustrative example devoted to the effectiveness comparison of the proposed approaches in the confidence estimation of the elementary models in the GMDH networks. Section 5.6 gives an example of the application of the GMDH neural-model based approach to the robust fault detection of an valve actuator in the Lublin Sugar Factory in Poland. It is worth to emphasis that the results described in this chapter are based on [71, 152, 153, 158, 159, 160, 161, 162].

5.2 Confidence Estimation of GMDH Neural Model

5.2.1 Least-Mean Square Method

Designing robust fault detection system requires the description of the model uncertainty. In particular, the problem is to obtain the parameters estimate vector $\hat{\boldsymbol{p}}_{n,k}^{(l)}$, as well as the associated parameter uncertainty required to design of the adaptive threshold. Indeed, since the bounds of the admissible changes of the system behaviour in the normal operating mode are known, violating these bounds by the system means that it is in a faulty mode. In order to obtain the GMDH neural model uncertainty it is necessary to calculate the confidence region of the model output for each neuron (3.2) in the GMDH network. The knowledge regarding the parameters estimate of the elementary model obtained with the application of the LMS method:

$$\hat{\boldsymbol{p}} = \left[\boldsymbol{R}^T\boldsymbol{R}\right]^{-1}\boldsymbol{R}^T\boldsymbol{Y}, \tag{5.1}$$

allows to obtain the confidence region of the parameters $\boldsymbol{p}$ with the confidence level of $(1-\alpha)$:

$$\hat{p}_i - t_{\alpha,n_{\mathcal{D}}-n_p-1}\sqrt{\hat{\sigma}^2 c_{ii}} < p_i < \hat{p}_i + t_{\alpha,n_{\mathcal{D}}-n_p-1}\sqrt{\hat{\sigma}^2 c_{ii}}, \quad i = 1, \ldots, n_p, \tag{5.2}$$

where c_{ii} represents i-th diagonal element of the matrix $C = (\boldsymbol{R}^T\boldsymbol{R})^{-1}$, $t_{\alpha,n_{\mathcal{D}}-n_p-1}$ represents $(1-\alpha)$ -th order quantel of a random variable which has the T-Student distribution with $(n_{\mathcal{D}} - n_p - 1)$ degrees of freedom, $\hat{\sigma}^2$ represents a variance of the random variable defined as a difference of the system output and its estimate $y_k - \hat{y}_k$:

$$\hat{\sigma}^2 = \frac{1}{n_{\mathcal{D}} - n_p - 1}\sum_{k=1}^{n_{\mathcal{D}}}(y_k - \hat{y}_k)^2 = \frac{\boldsymbol{Y}^T\boldsymbol{Y} - \hat{\boldsymbol{p}}^T\boldsymbol{R}^T\boldsymbol{Y}}{n_{\mathcal{D}} - n_p - 1}. \tag{5.3}$$

Finally, the confidence region of the parameters $\boldsymbol{p}$ with the confidence level of $(1-\alpha)$ can be defined as follows:

$$(\hat{\boldsymbol{p}} - \boldsymbol{p})^T\boldsymbol{R}^T\boldsymbol{R}(\hat{\boldsymbol{p}} - \boldsymbol{p}) \leq (n_p + 1)\hat{\sigma}^2 F^{n_p+1}_{\alpha,n_{\mathcal{D}}-n_p-1}, \tag{5.4}$$

where $F^{n_p+1}_{\alpha,n_{\mathcal{D}}-n_p-1}$ is $(1-\alpha)$-th order quantel of the random variable which has the Snedecor's F-Distribution with $(n_{\mathcal{D}} - n_p - 1)$ and $(n_p + 1)$ degrees of freedom.

In order to obtain the $(1-\alpha)$ confidence interval for the following neuron output:

$$\hat{y}_k = \boldsymbol{r}_k^T\hat{\boldsymbol{p}}, \tag{5.5}$$

it is necessary to assume that $\hat{y}_k$ is a random variable, which has a Gaussian distribution. Then the expected value is:

$$\mathcal{E}\left[y_k\right] = \mathcal{E}\left[\boldsymbol{r}_k^T\hat{\boldsymbol{p}}\right] = \boldsymbol{r}_k^T\mathcal{E}\left[\hat{\boldsymbol{p}}\right] = \boldsymbol{r}_k^T\boldsymbol{p}, \tag{5.6}$$

and the variance is:

$$\begin{aligned} \operatorname{var}\left[y_k\right] &= \mathcal{E}\left[(\hat{y}_k - \mathcal{E}\left[\hat{y}_k\right])(\hat{y}_k - \mathcal{E}\left[\hat{y}_k\right])^T\right] = & (5.7)\\ &= \mathcal{E}\left[\boldsymbol{r}_k^T(\hat{\boldsymbol{p}} - \boldsymbol{p})(\hat{\boldsymbol{p}} - \boldsymbol{p})^T\boldsymbol{r}_k\right] = & (5.8)\\ &= \boldsymbol{r}_k^T\mathcal{E}\left[(\hat{\boldsymbol{p}} - \boldsymbol{p})(\hat{\boldsymbol{p}} - \boldsymbol{p})^T\right]\boldsymbol{r}_k. \end{aligned}$$

Taking into consideration, that:

$$\mathcal{E}\left[(\hat{\boldsymbol{p}} - \boldsymbol{p})(\hat{\boldsymbol{p}} - \boldsymbol{p})^T\right] = \operatorname{cov}\left[\boldsymbol{p}\right] = \left(\boldsymbol{R}^T\boldsymbol{R}\right)^{-1}\sigma^2, \tag{5.9}$$

the relation (5.7) has the following form:

$$\operatorname{var}\left[y_k\right] = \boldsymbol{r}_k^T\left(\boldsymbol{R}^T\boldsymbol{R}\right)^{-1}\boldsymbol{r}_k\sigma^2. \tag{5.10}$$

Finally, the $(1-\alpha)$ confidence interval for the neuron output (5.5) is defined as:

$$\begin{aligned}\hat{y}_k - t_{\alpha,n_D-n_p-1}\sqrt{\hat{\sigma}^2\boldsymbol{r}_k^T\left(\boldsymbol{R}^T\boldsymbol{R}\right)^{-1}\boldsymbol{r}_k} < \boldsymbol{r}_k^T\boldsymbol{p} < \\ \hat{y}_k + t_{\alpha,n_D-n_p-1}\sqrt{\hat{\sigma}^2\boldsymbol{r}_k^T\left(\boldsymbol{R}^T\boldsymbol{R}\right)^{-1}\boldsymbol{r}_k}.\end{aligned} \tag{5.11}$$

The neuron output uncertainty defined by (5.11) is presented in Fig. 5.1.

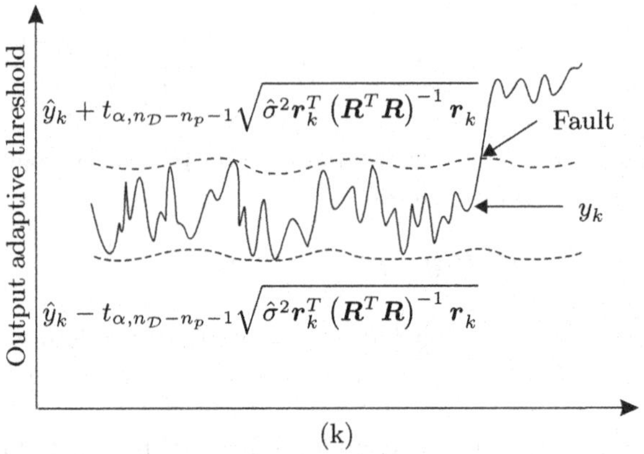

Fig. 5.1. Output adaptive threshold obtained via the LMS

Based on the each neuron output uncertainty (5.11) it is relatively easy to obtain the uncertainty of the whole GMDH neural network. Unfortunately, the application of the proposed approach in the real fault detection task may lead to the false alarms and undetected faults. It follows from the fact that the condition (2.101) concerning noise may not have been fulfilled in the real fault detection tasks. Moreover, the error in the regressor of the neuron, i.e., an uncertain regressor, can additionally degrade solutions obtained by the LMS method. For this reason it is necessary to employ a more reliable and more robust methodology.

5.2.2 Bounded-Error Approach

The problems detailed in the previous subsection show that there is a need for the application of a parameters estimation method different than the LMS method in order to obtain the adaptive threshold which can be applied in the robust fault detection. Such a method should be also easily adaptable to the case of an uncertain regressor and it should overcome all remaining difficulties mentioned in Sect. 2.6. The subsequent part of this section gives an outline of such a method which is based on the BEA. The methodology described in Sect. 3.3.2 makes it possible to obtain the parameters estimate $\hat{\boldsymbol{p}}$ and associated feasible parameters set $\mathbb{P}$ which can be used to obtain model output uncertainty, i.e., the interval in which the "true" model output $\hat{y}_k$ can be found. This kind of knowledge makes it possible to obtain an adaptive threshold [20], and hence to develop a fault diagnosis scheme that is robust to model uncertainty.

Let $\mathbb{V}$ be the set of all vertices $\boldsymbol{p}^i$, $i = 1, \ldots, n_v$, describing the feasible parameters set $\mathbb{P}$ (cf. (3.20)). If there is no error in the regressor, then the problem of determining the model output uncertainty can be solved as follows:

$$\boldsymbol{r}_k^T \hat{\boldsymbol{p}}_k^m \leq \boldsymbol{r}_k^T \boldsymbol{p} \leq \boldsymbol{r}_k^T \hat{\boldsymbol{p}}_k^M, \tag{5.12}$$

where:

$$\hat{\boldsymbol{p}}_k^m = \arg\min_{\boldsymbol{p}\in\mathbb{V}} \ \boldsymbol{r}_k^T \boldsymbol{p}, \tag{5.13}$$

$$\hat{\boldsymbol{p}}_k^M = \arg\max_{\boldsymbol{p}\in\mathbb{V}} \boldsymbol{r}_k^T \boldsymbol{p}. \tag{5.14}$$

The computation of (5.13) and (5.14) is realised by multiplying the parameter vectors corresponding to all vertices belonging to $\mathbb{V}$ by the regressor $\boldsymbol{r}_k^T$ what results in obtaining the model output uncertainty interval depicted in Fig. 5.2.

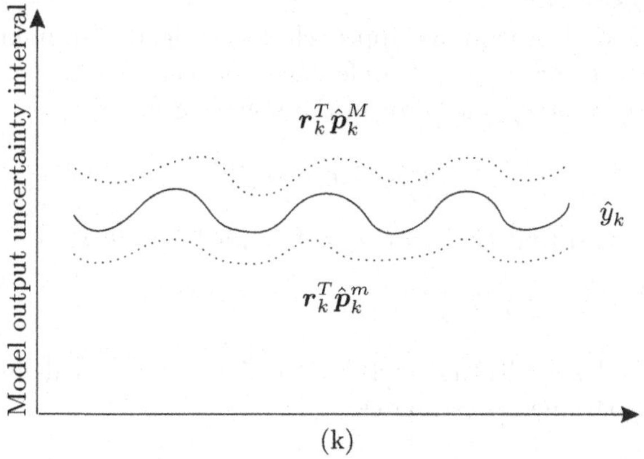

Fig. 5.2. Model output uncertainty obtained via the BEA for the error-free regressor

Since (5.12) describes the neuron output uncertainty interval, the system output adaptive threshold satisfies:

$$\boldsymbol{r}_k^T \hat{\boldsymbol{p}}_k^m + \varepsilon_k^m \leq y_k \leq \boldsymbol{r}_k^T \hat{\boldsymbol{p}}_k^M + \varepsilon_k^M. \tag{5.15}$$

and it is presented in Fig. 5.3. A more general case of (5.15) for neurons with a non-linear activation function is considered in Sect. 5.3.

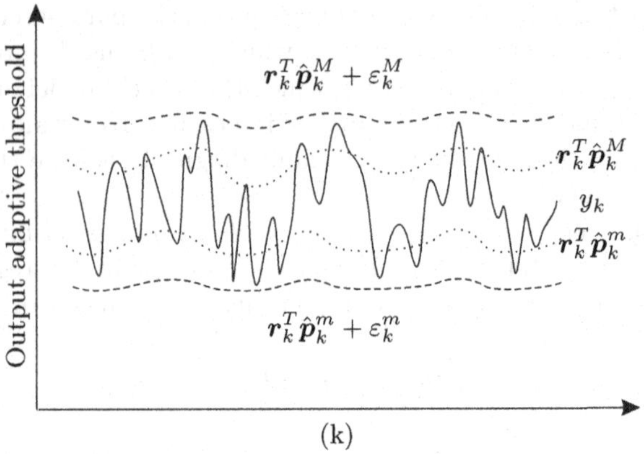

Fig. 5.3. Output adaptive threshold obtained via the BEA for the error-free regressor

As it has been already mentioned, the neurons in the l-th ($l > 1$) layer are fed with the outputs of the neurons from the $(l-1)$-th layer. Since (5.12) describes the model output uncertainty, the parameters of the neurons in layers have to be obtained with an approach that solves the problem of an uncertain regressor [15].

In order to modify the above approach, let us denote an unknown "true" value of the regressor $\boldsymbol{r}_{n,k}$ by a difference between the known (measured) value of the regressor $\boldsymbol{r}_k$ and error in the regressor $\boldsymbol{e}_k$:

$$\boldsymbol{r}_{n,k} = \boldsymbol{r}_k - \boldsymbol{e}_k, \tag{5.16}$$

where it is assumed that the error $\boldsymbol{e}_k$ is bounded as follows:

$$e_{i,k}^m \leq e_{i,k} \leq e_{i,k}^M, \quad i = 1, \ldots, n_p. \tag{5.17}$$

Using (3.16) and substituting (5.16) into (5.17) one can define the space containing the parameters estimates:

$$\varepsilon_k^m - \boldsymbol{e}_k^T\boldsymbol{p} \leq y_k - \boldsymbol{r}_k^T\boldsymbol{p} \leq \varepsilon_k^M - \boldsymbol{e}_k^T\boldsymbol{p}. \tag{5.18}$$

Unfortunately, for the purpose of the parameters estimation it is not enough to introduce (5.16) into (5.17). Indeed, the bounds of (5.18) depend also on the sign of each p_i. It is possible to obtain directly these signs only for models which parameters have a physical meaning [163]. For models such as the GMDH neural networks it is rather impossible. In [15], the authors proposed some heuristic techniques, but these drastically complicate the problem (5.18) and do not to guarantee that these signs will be obtained properly. Bearing in mind the fact that the neuron (3.2) contains only a few parameters, it is possible to replace them by:

$$p_i = p_i' - p_i'', \quad p_i', p_i'' \geq 0, \quad i = 1, \ldots, n_p. \tag{5.19}$$

Although the above solution is very simple, it doubles the number of parameters, i.e., instead of estimating n_p parameters it is necessary to do so for $2n_p$ parameters. In spite of that, this technique is very popular and widely used in the literature [15, 164]. Due to the above solution, (5.18) can be modified as follows:

$$\begin{aligned} &\varepsilon_k^m - \left(\boldsymbol{e}_k^M\right)^T \boldsymbol{p}' + \left(\boldsymbol{e}_k^m\right)^T \boldsymbol{p}'' \leq y_k - \boldsymbol{r}_k^T(\boldsymbol{p}' - \boldsymbol{p}'') \leq \\ &\varepsilon_k^M - \left(\boldsymbol{e}_k^m\right)^T \boldsymbol{p}' + \left(\boldsymbol{e}_k^M\right)^T \boldsymbol{p}''. \end{aligned} \tag{5.20}$$

This transformation makes it possible to employ, with a minor modification, the approach described in Sect. 3.3.2. The difference is that the algorithm processes each constraint associated with a pair of hyperplanes defined with (5.20) separately. The reason for such a modification is that the hyperplanes are not parallel [147]. Figure 5.4 presents feasible parameters set obtained by the intersection of the strips (5.20) for $n_p = 2$.

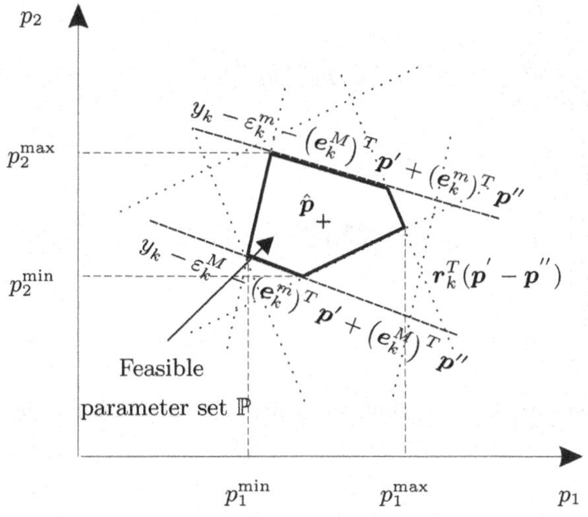

Fig. 5.4. Feasible parameters set obtained via the BEA for the error in the regressor

The proposed modification of the BEA makes it possible to estimate the parameter vectors of the neurons from the l-th, $l > 1$ layers. In the case of an error in the regressor, using (5.20), it can be shown that the model output uncertainty has the following form:

$$\hat{y}_k^m(\hat{\boldsymbol{p}}_k^{'m}, \hat{\boldsymbol{p}}_k^{''m}) \leq \boldsymbol{r}_{n,k}^T \boldsymbol{p} \leq \hat{y}_k^M(\hat{\boldsymbol{p}}_k^{'M}, \hat{\boldsymbol{p}}_k^{''M}), \tag{5.21}$$

where:

$$\hat{y}_k^m(\hat{\boldsymbol{p}}_k^{'m},\hat{\boldsymbol{p}}_k^{''m}) = \left(\boldsymbol{r}_k - \boldsymbol{e}_k^M\right)^T \boldsymbol{p}_k^{'m} + \left(\boldsymbol{e}_k^m - \boldsymbol{r}_k\right)^T \boldsymbol{p}_k^{''m}, \tag{5.22}$$

and

$$\hat{y}_k^M(\hat{\boldsymbol{p}}_k^{'M},\hat{\boldsymbol{p}}_k^{''M}) = \left(\boldsymbol{r}_k - \boldsymbol{e}_k^m\right)^T \boldsymbol{p}_k^{'M} + \left(\boldsymbol{e}_k^M - \boldsymbol{r}_k\right)^T \boldsymbol{p}_k^{''M}, \tag{5.23}$$

where:

$$\left(\hat{\boldsymbol{p}}_k^{'m},\hat{\boldsymbol{p}}_k^{''m}\right) = \arg\min_{(\boldsymbol{p}',\boldsymbol{p}'')\in\mathbb{V}} \hat{y}_k^m(\boldsymbol{p}',\boldsymbol{p}_k''), \tag{5.24}$$

and

$$\left(\hat{\boldsymbol{p}}_k^{'M},\hat{\boldsymbol{p}}_k^{''M}\right) = \arg\max_{(\boldsymbol{p}',\boldsymbol{p}'')\in\mathbb{V}} \hat{y}_k^M(\boldsymbol{p}',\boldsymbol{p}_k''). \tag{5.25}$$

The model output uncertainty interval calculated with taking into account error into regressor case is depicted in Fig. 5.5.

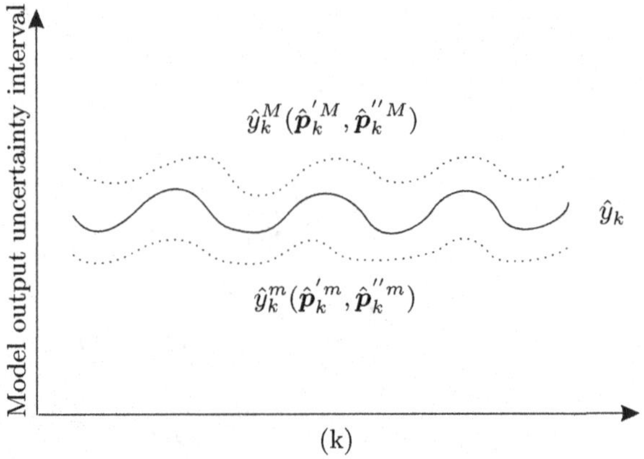

Fig. 5.5. Model output uncertainty obtained via the BEA for the error in the regressor

Using the interval (5.21) it is possible to obtain the system output adaptive threshold which can be used in the robust fault detection tasks:

$$\hat{y}_k^m(\hat{\boldsymbol{p}}_k^{'m},\hat{\boldsymbol{p}}_k^{''m}) + \varepsilon_k^m \leq y_k \leq \hat{y}_k^M(\hat{\boldsymbol{p}}_k^{'M},\hat{\boldsymbol{p}}_k^{''M}) + \varepsilon_k^M. \tag{5.26}$$

The principle of the fault detection with the developed adaptive threshold is shown in Fig. 5.6. The fault is detected when the response of the diagnosed system crosses the output adaptive threshold (5.26).

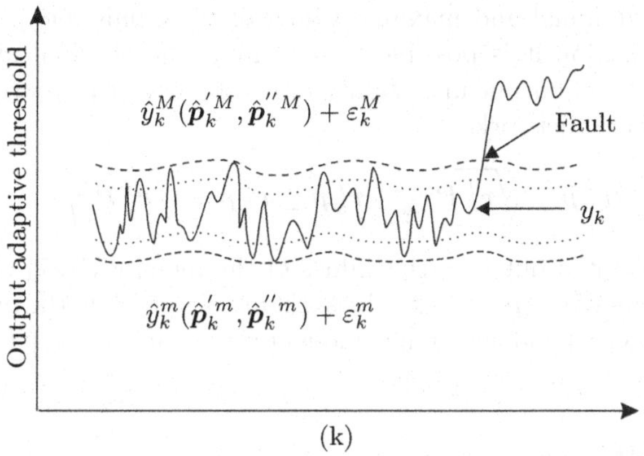

Fig. 5.6. Output adaptive threshold obtained via the BEA for the error in the regresor

5.2.3 Outer Bounding Ellipsoid Algorithm

In Sect. 3.3.3 the methodology of the calculation of the parameters estimate $\hat{\boldsymbol{p}}$, as well as the associated parameter uncertainty $\mathbb{E}$ with the application of the OBE algorithm required to design robust fault detection system is presented. This knowledge allows to obtain the model output uncertainty interval containing the "true" model output $\hat{y}_k$. The range of the confidence interval of the partial model output depends on the size and the orientation of the elipsoid which defines the admissible parameter set $\mathbb{E}$ (cf. Fig. 5.7).

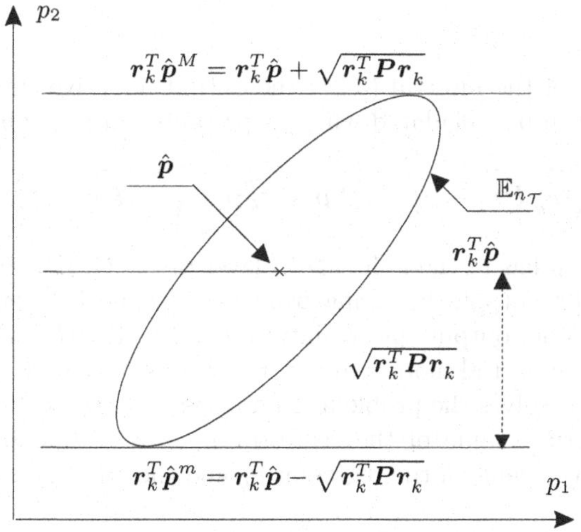

Fig. 5.7. Relation between the ellipsoid and the model output uncertainty

Taking the minimal and maximal values of the admissible parameter set $\mathbb{E}$ into consideration it is possible to determine the minimal and maximal values of the model output uncertainty interval for the each partial model of the GMDH neural network:

$$\boldsymbol{r}_k^T\hat{\boldsymbol{p}} - \sqrt{\boldsymbol{r}_k^T\boldsymbol{P}\boldsymbol{r}_k} \leq \boldsymbol{r}_k^T\boldsymbol{p} \leq \boldsymbol{r}_k^T\hat{\boldsymbol{p}} + \sqrt{\boldsymbol{r}_k^T\boldsymbol{P}\boldsymbol{r}_k}. \tag{5.27}$$

It should be pointed out that the values of the interval (5.27) change along with the changes of the regressor values. Figure 5.8 presents the model output uncertainty interval and the "true" model output for the error-free regressor case.

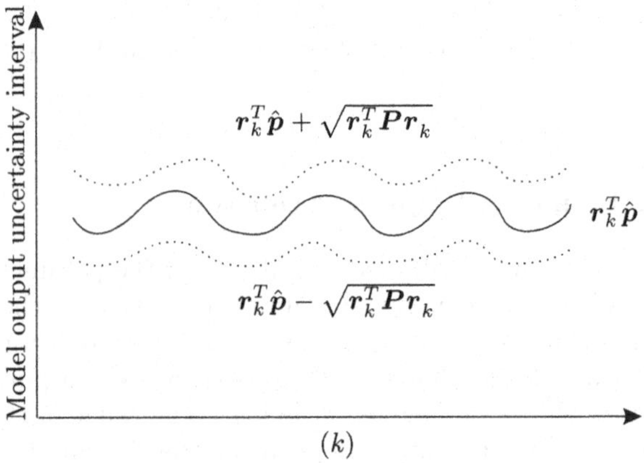

Fig. 5.8. Model output uncertainty obtained via the OBE for the error-free regressor

On the basis of the interval (5.27) the output adaptive threshold for the error-free case can be calculated which is presented in Fig. 5.9.

$$\boldsymbol{r}_k^T\hat{\boldsymbol{p}} - \sqrt{\boldsymbol{r}_k^T\boldsymbol{P}\boldsymbol{r}_k} + \varepsilon_k^m \leq \boldsymbol{r}_k^T\boldsymbol{p} \leq \boldsymbol{r}_k^T\hat{\boldsymbol{p}} + \sqrt{\boldsymbol{r}_k^T\boldsymbol{P}\boldsymbol{r}_k} + \varepsilon_k^M. \tag{5.28}$$

The partial models in the l-th ($l > 1$) layer of the GMDH neural network are based on the outputs incoming from the $(l-1)$-th layer. Since (5.27) describes the model output uncertainty interval in the $(l-1)$-th layer, the parameters of the partial models in the next layers have to be obtained with an approach that solves the problem of an uncertain regressor. Let us denote an unknown "true" value of the regressor $\boldsymbol{r}_{n,k}$ by a difference between a known (measured) value of the regressor $\boldsymbol{r}_k$ and the error in the regressor $\boldsymbol{e}_k$:

$$\boldsymbol{r}_{n,k} = \boldsymbol{r}_k - \boldsymbol{e}_k, \tag{5.29}$$

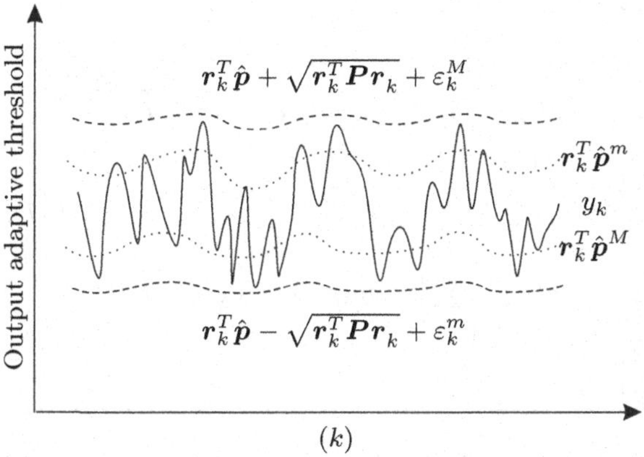

Fig. 5.9. Output adaptive threshold obtained via the OBE for error-free regressor

where the regressor error $\boldsymbol{e}_k$ is bounded as follows:

$$-\epsilon_i \leq e_{i,k} \leq \epsilon_i, \quad i = 1, \ldots, n_p. \tag{5.30}$$

Substituting (5.29) into (5.27) it can be shown that the partial models output uncertainty interval have the following form: (cf. Fig. 5.10):

$$\hat{y}_k^m \leq \boldsymbol{r}_k^T \boldsymbol{p} \leq \hat{y}_k^M, \tag{5.31}$$

where:

$$\hat{y}_k^m = \boldsymbol{r}_{n,k}^T \hat{\boldsymbol{p}} + \boldsymbol{e}_k^T \hat{\boldsymbol{p}} - \sqrt{(\boldsymbol{r}_{n,k} + \boldsymbol{e}_k)^T \boldsymbol{P} (\boldsymbol{r}_{n,k} + \boldsymbol{e}_k)}, \tag{5.32}$$

and

$$\hat{y}_k^M = \boldsymbol{r}_{n,k}^T \hat{\boldsymbol{p}} + \boldsymbol{e}_k^T \hat{\boldsymbol{p}} + \sqrt{(\boldsymbol{r}_{n,k} + \boldsymbol{e}_k)^T \boldsymbol{P} (\boldsymbol{r}_{n,k} + \boldsymbol{e}_k)}. \tag{5.33}$$

In order to obtain the final form of the expression (5.31) it is necessary to take into consideration the bounds of the regressor error (5.30) in the expressions (5.32) and (5.33):

$$\hat{y}_k^m = \boldsymbol{r}_{n,k}^T \hat{\boldsymbol{p}} + \sum_{i=1}^{n_p} \mathrm{sgn}(\hat{p}_i)\hat{p}_i \epsilon_i - \sqrt{\bar{\boldsymbol{r}}_{n,k}^T \boldsymbol{P} \bar{\boldsymbol{r}}_{n,k}}, \tag{5.34}$$

and

$$\hat{y}_k^M = \boldsymbol{r}_{n,k}^T \hat{\boldsymbol{p}} + \sum_{i=1}^{n_p} \mathrm{sgn}(\hat{p}_i)\hat{p}_i \epsilon_i + \sqrt{\bar{\boldsymbol{r}}_{n,k}^T \boldsymbol{P} \bar{\boldsymbol{r}}_{n,k}}, \tag{5.35}$$

where:

$$\bar{r}_{n,i,k} = r_{n,i,k} + \mathrm{sgn}\left(r_{n,i,k}\right) \epsilon_i. \tag{5.36}$$

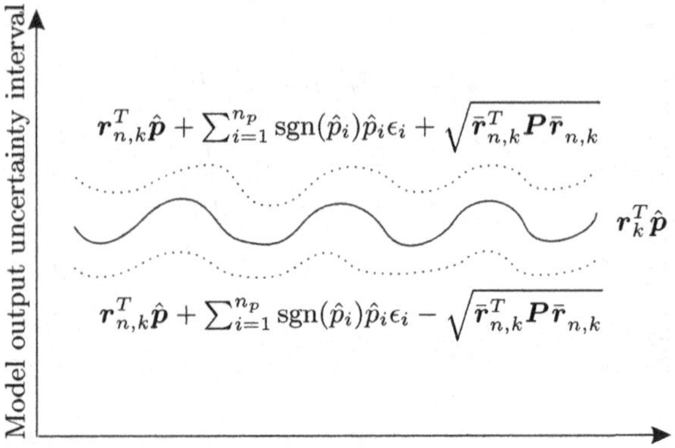

Fig. 5.10. Model output uncertainty obtained via the OBE for the error in the regressor

On the basis of the expressions (5.34)-(5.36) the output adaptive threshold can be defined:

$$\begin{aligned} &\boldsymbol{r}_{n,k}^T\hat{\boldsymbol{p}} + \sum_{i=1}^{n_p} \mathrm{sgn}(\hat{p}_i)\hat{p}_i\epsilon_i - \sqrt{\bar{\boldsymbol{r}}_{n,k}^T \boldsymbol{P} \bar{\boldsymbol{r}}_{n,k}} + \varepsilon_k^m \leq y_k \leq \\ &\boldsymbol{r}_{n,k}^T\hat{\boldsymbol{p}} + \sum_{i=1}^{n_p} \mathrm{sgn}(\hat{p}_i)\hat{p}_i\epsilon_i + \sqrt{\bar{\boldsymbol{r}}_{n,k}^T \boldsymbol{P} \bar{\boldsymbol{r}}_{n,k}} + \varepsilon_k^M. \end{aligned} \tag{5.37}$$

Above defined threshold presented in Fig. 5.11 should contain the system response in the fault-free mode.

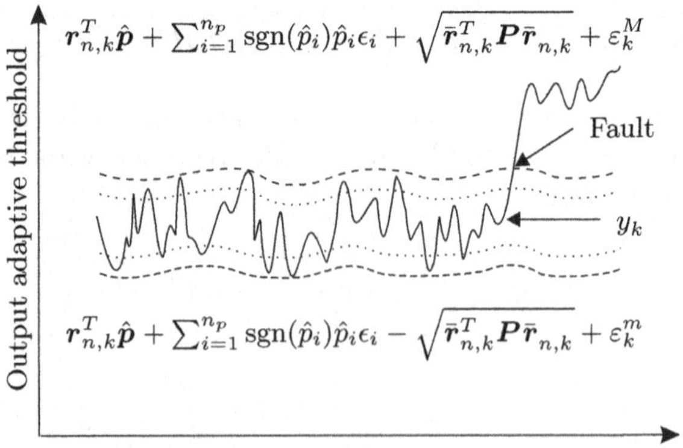

Fig. 5.11. Output adaptive threshold obtained via the OBE for the error in the regressor

5.2.4 Zonotope-Based Algorithm

The description of neurons uncertainty in the form of the zonotope $\mathbb{Z}$ (3.71) allows to obtain a GMDH neural model-based robust fault detection scheme. On the basis of (3.68)-(3.69) the neuron output uncertainty interval can be defined as follows:

$$\boldsymbol{r}_k^T\hat{\boldsymbol{p}} - \|\mathbb{P}^T\boldsymbol{r}_k\|_1 \leq \boldsymbol{r}_k^T\hat{\boldsymbol{p}} \leq \boldsymbol{r}_k^T\hat{\boldsymbol{p}} + \|\mathbb{P}^T\boldsymbol{r}_k\|_1. \tag{5.38}$$

Figure 5.12 presents the model output uncertainty interval containing the model response.

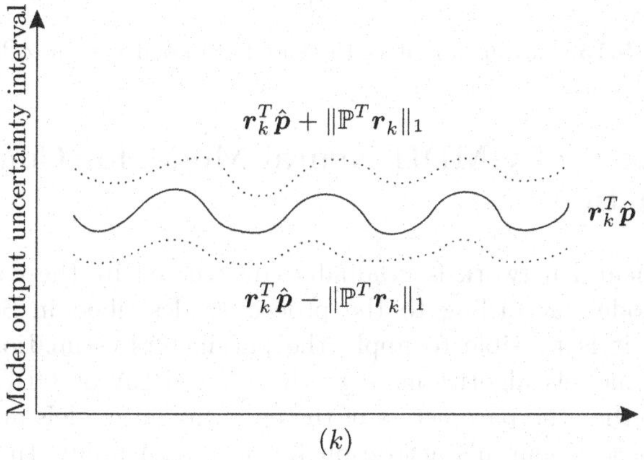

Fig. 5.12. Model output uncertainty obtained via the ZBE for the error-free regressor

Additionally, as the measurement of the diagnosed system responses $\boldsymbol{y}_k$ is corrupted by the noise, it is necessary to add the boundary values of the output error to the model output uncertainty interval. In the proposed approach these values are overbounded by σ according to the relation (3.61). Since (5.38) describes the neuron output uncertainty interval, the system output adaptive threshold satisfies:

$$\boldsymbol{r}_k^T\hat{\boldsymbol{p}} - \|\mathbb{P}^T\boldsymbol{r}_k\|_1 - \sigma \leq \boldsymbol{y}_k \leq \boldsymbol{r}_k^T\hat{\boldsymbol{p}} + \|\mathbb{P}^T\boldsymbol{r}_k\|_1 + \sigma, \tag{5.39}$$

and is presented in Fig. 5.13. Defined by (5.39) the output adaptive threshold should contain the real system response in the fault free mode.

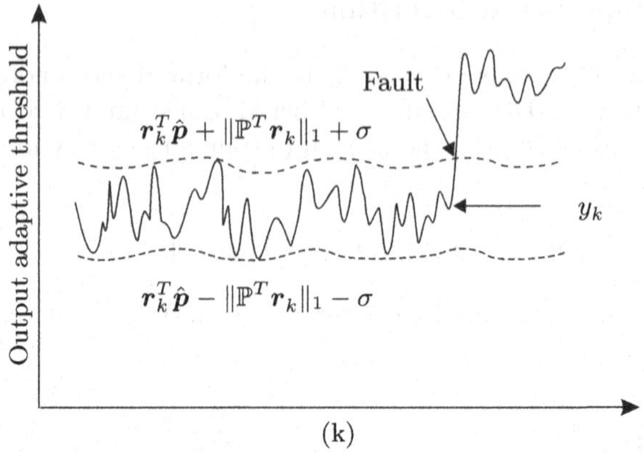

Fig. 5.13. Output adaptive threshold obtained via the ZBA

5.3 Synthesis of GMDH Neural Model for Confidence Estimation

The GMDH neural network is gradually constructed by the connection of the partial models according to the procedure described in Sect. 2.4. In this approach it is possible to apply the parameters estimation of linear-in-parameters models algorithms, e.g., the LMS, BEA or OBE. It follows from the facts that the parameters of the each partial models are estimated separately and the neuron's activation function $f(\cdot)$ fulfills the conditions (2.42-2.43).

On the beginning, it is necessary to adapt the BEA and OBE algorithms to the parameters estimation of the partial models with the non-linear activation function $f(\cdot)$. In order to avoid the noise additivity problem described in Sect. 2.6 with the application of the BEA, OBE and ZBA algorithms, it is necessary to transform the following relation:

$$\varepsilon_k^m \le y_k - f\left(\left(\boldsymbol{r}_{n,k}^{(l)}\right)^T \boldsymbol{p}_n^{(l)}\right) \le \varepsilon_k^M \tag{5.40}$$

using $f^{-1}(\cdot)$, and hence:

$$f^{-1}\left(y_k - \varepsilon_k^M\right) \le \left(\boldsymbol{r}_{n,k}^{(l)}\right)^T \boldsymbol{p}_n^{(l)} \le f^{-1}\left(y_k - \varepsilon_k^m\right). \tag{5.41}$$

The transformation (5.41) is appropriate if the conditions (2.42) and (2.43) concerning the properties of the non-linear activation function $f(\cdot)$ are fulfilled. For the parameters and confidence estimation of the partial models in the first layer of the GMDH neural network, the BEA, OBE and ZBA algorithms not taking into account the error in the regressor can be applied. The boundary values of the neurons output uncertainty interval are obtained on the basis of the inequalities (5.21), (5.31) and (5.38) for the BEA, OBE and ZBA algorithms, respectively. The outputs of the selected partial models become the inputs to other partial models in the next layer. Because of partial model output is known with some confidence defined by the model output uncertainty interval, therefore during the design procedure of the neurons from the second and the subsequent layers of the GMDH neural network an error in the regressor has to be considered.

In the case of the BEA the application of (5.12) in the first layer and (5.21) in the subsequent ones, allows to obtain the bounds of the neuron output and the bounds of the regressor error (5.17), whilst the known value of the regressor should be computed by using the parameters estimates $\hat{\boldsymbol{p}}_n^{(l)}$. Similarly, in the case of the OBE algorithm, using (5.27) in the first layer and (5.31) in the subsequent ones, it is possible to obtain the bounds of the partial, models outputs, and at the same time are the bounds (5.30) of the regressor error of the partial models in the next layer.

The presented parameters estimation approaches make it possible to obtain the partial models uncertainty. Furthermore, they reduce the contribution of the parameters estimates inaccuracy to the model uncertainty. Nevertheless, the model uncertainty is not completely eliminated. It should be pointed out that the processing errors of the partial models, which are described by the model output uncertainty (5.31) for the OBE and (5.21) for the BEA, can be propagated and accumulated during the introduction of new layers (cf. Fig. 5.14).

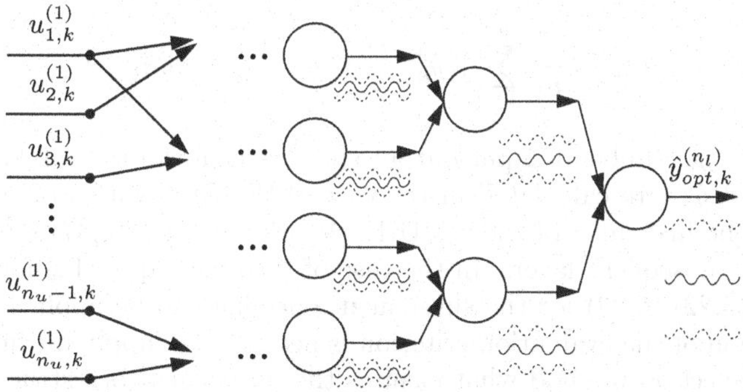

Fig. 5.14. Accumulation of model output uncertainty

The scale of this unfavourable phenomenon depends on the applied evaluation criterion in the partial models and selection method. As it is mentioned in Sect. 2.4, the selection methods in the GMDH neural networks plays a role of a mechanism of the structural optimisation. During the selection, partial models, which have too large value of the defined evaluation criterion, are rejected on the basis of chosen selection methods. Unfortunately, as it was mentioned in Sect. 2.6, the application of the classical evaluation criteria like the AIC and FPE [70, 118] during the network synthesis may lead to the selection of an inappropriate structure of the GMDH neural network. It follows from the fact that the above criteria does not take into account the modelling uncertainty. In this way, the partial models with the small values of the classical evaluation criteria $Q_{\mathcal{V}}$ but with large uncertainty can be obtained (cf. Fig. 5.15).

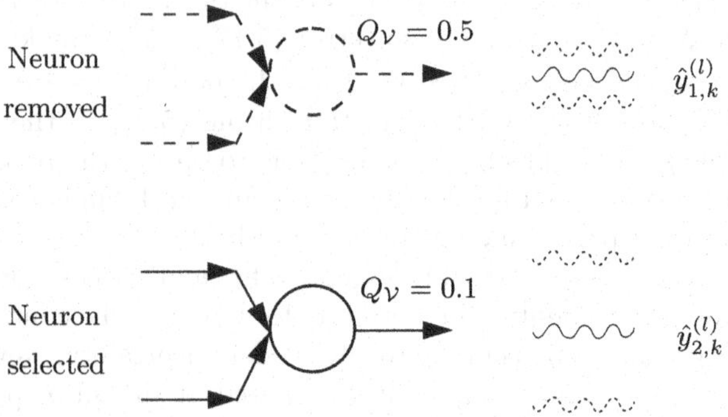

Fig. 5.15. Problem of an incorrect selection of a neurons

In order to overcome this difficulty, the following evaluation criterion can be used:

$$Q_{\mathcal{V}} = \frac{1}{n_{\mathcal{V}}} \sum_{k=1}^{n_{\mathcal{V}}} \left| \left(\hat{y}_k^M + \varepsilon_k^M\right) - \left(\hat{y}_k^m + \varepsilon_k^m\right) \right|, \tag{5.42}$$

where $n_{\mathcal{V}}$ is the number of input-output measurements for the validation data set, $\hat{y}_k^M$ and $\hat{y}_k^m$ are calculated in the case of the BEA with (5.27) for each neuron in the first layer of the GMDH network and with (5.31) for the neurons in the subsequent layers. In the case of the OBE eq. (5.12) for the first layer and (5.22)-(5.23) for the subsequent ones have to be applied. Besides, the definition of the evaluation criterion is necessary to apply during the appropriate selection method what ensures the proper network structure. The selection methods described in Sect. 2.4 have several disadvantages which can lead to large uncertainty of the neural model. In order to overcome these difficulties the method based on the SSM can be applied (cf. Sect. 2.4). Finally, the neuron in the last layer that gives the smallest processing error (5.42) constitutes the output of the GMDH neural network, while the model output uncertainty of this neuron is used for the calculation of the output adaptive threshold that can be employed for robust fault detection.

5.4 Further Improvement of GMDH Neural Model

One of the main advantages of the GMDH neural networks is that the BEA for linear systems can be applied to estimate the parameters of each neuron. This is possible because the parameter vectors of the neurons are estimated independently. The application of this technique implies that the parameter vectors are obtained in an optimal way, i.e., there is no linearisation etc. However, the optimality should be perceived as a only local one. This means that the parameter vector associated with a neuron is optimal for this particular neuron. On the other hand, this parameter vector may not be optimal from the point of view of the entire network. Such circumstances rise the need for the retraining of the GMDH neural network after the automatic selection of the model structure.

Assuming that the GMDH neural network can be written in the following form:

$$\hat{y} \underline{g} \left(\hat{\boldsymbol{p}}_1^{(1)}, \ldots \hat{\boldsymbol{p}}_{n_y}^{(1)}, \ldots \hat{\boldsymbol{p}}_1^{(L)}, \ldots \hat{\boldsymbol{p}}_{n_{Ly}}^{(L)}, \boldsymbol{u}\right), \tag{5.43}$$

where $g(\cdot)$ stands for the neural network structure obtained with the GMDH approach, L is the number of layers, n_{iy} is the number of neurons in the i-th layer. After the procedure of designing the GMDH neural network, described in Sect. 5.3, the following sequence of estimates is obtained: $\hat{\boldsymbol{p}}_1^{(1)}, \ldots \hat{\boldsymbol{p}}_{n_y}^{(1)}, \ldots \hat{\boldsymbol{p}}_1^{(L)}, \ldots \hat{\boldsymbol{p}}_{n_y}^{(L)}$. With each of these estimates, a feasible parameters set is associated, i.e.:

$$\boldsymbol{A}_{i,j}\boldsymbol{p}_i^{(j)} \leq \boldsymbol{b}_{i,j}, \quad i = 1, \ldots, n_{jy},\, j = 1, \ldots, L, \tag{5.44}$$

where the matrix $\boldsymbol{A}_{i,j}$ and vector $\boldsymbol{b}_{i,j}$ are used for describing the feasible parameters sets of each neuron of (5.43). These feasible parameters sets are known after the automatic selection of the model structure and estimation of its parameters. It enables to formulate the parameters estimation task in a global sense. Indeed, it can be defined as a constrained optimisation problem of:

$$Q_{\mathcal{T}} = \frac{1}{n_{\mathcal{T}}} \sum_{k=1}^{n_{\mathcal{T}}} |y_k - \hat{y}_k|, \tag{5.45}$$

and

$$\hat{y}_k = g\left(\hat{\boldsymbol{p}}_1^{(1)}, \ldots \hat{\boldsymbol{p}}_{n_y}^{(1)}, \ldots \hat{\boldsymbol{p}}_1^{(L)}, \ldots \hat{\boldsymbol{p}}_{n_y}^{(L)}, \boldsymbol{u}_k\right), \tag{5.46}$$

where the constraints are given by (5.44). The solution of (5.45) can be obtained with the optimisation techniques related with non-linear l_1 optimisation [16] as well as specialised evolutionary algorithms [165]. Based on numerous computer experiments, it has been found that the simple ARS algorithm [16] is an especially well-suited for that purpose. Apart from its simplicity, the algorithm decreases the chance of getting stuck in a local optimum and hence, it may give a global minimum of (5.45).

5.5 An Illustrative Example – Confidence Estimation of Neurons

The purpose of the present section is to show the effectiveness of the proposed approaches based on the LMS, BEA, OBE and Zonotpoe-based algorithms in the task of confidence estimation and calculation of the adaptive thresholds for the neurons in the GMDH network. The obtained feasible parameters sets in Sects. 3.3.1 - 3.3.4 describe the parametrical uncertainty of the each neuron in the GMDH network. On the basis of the obtained matrixes $\mathbb{P}$ it is possible to calculate the neuron uncertainty in the form of the confidence interval of the model output according to the approaches described in Sects. 5.2.1-5.2.4. Taking into account these results and the disturbances ε the calculation of the output adaptive thresholds, which can be applied to robust fault diagnosis is enabled. For all methods the output adaptive thresholds are calculated for the following common validation signal:

$$u_{1,k} = \sin(2\pi(k)/250) \quad \text{for} \quad k = 1, \ldots, 100,$$

$$u_{2,k} = 0.8\sin(2\pi(k/250)) + 0.2\sin(2\pi(k/25)) \quad \text{for} \quad k = 101, \ldots, 200.$$

On the basis of the calculated matrixes $\mathbb{P}$ for all estimation methods describing the feasible parameters sets, which are presented in Figs. 3.7-3.10 and in Tab. 3.1, the output adaptive thresholds are calculated and depicted in Figs. 5.16-5.19. In the case of the LMS method the system output adaptive threshold is calculated with the application of expression (5.11) and together with system output is depicted in Fig. 5.16. The results obtained with the LMS method indicate that the system output adaptive threshold does not contain the system output calculated based on the nominal parameters $\boldsymbol{p}$.

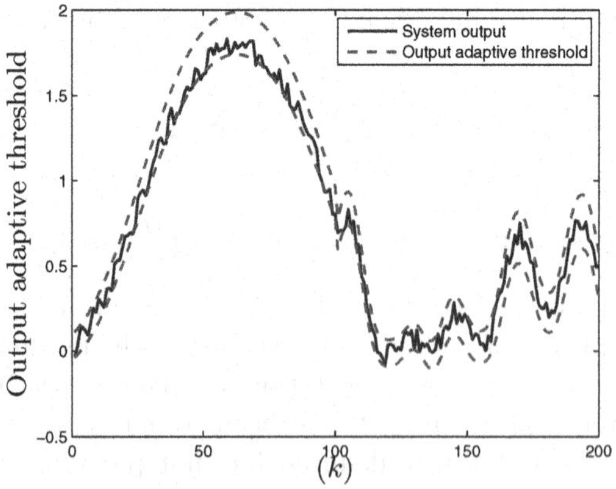

Fig. 5.16. Output adaptive threshold obtained via the LMS

This result is consistent with the expectations because the parameters estimate obtained with the application of the LMS method is biased and the feasible parameters set does not contain the nominal parameters (cf. Fig 3.7).

The opposite result is obtain for the BEA where the parameters estimate is unbiased and thus, the calculated feasible parameters set contains the nominal parameters (cf. Fig. 3.8). Hence, the output adaptive threshold presented in Fig. 5.17 and calculated according (5.26) includes the nominal system response. It should be also underlined that the output adaptive threshold obtained with the application of the BEA is the narrowest what results from the smallest region of the feasible parameters set.

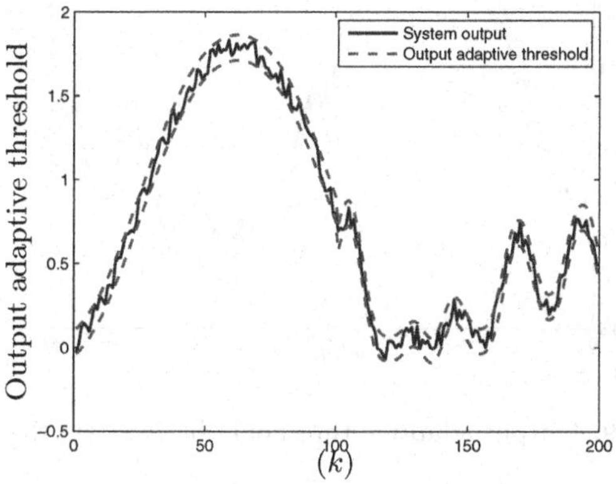

Fig. 5.17. Output adaptive threshold obtained via the BEA

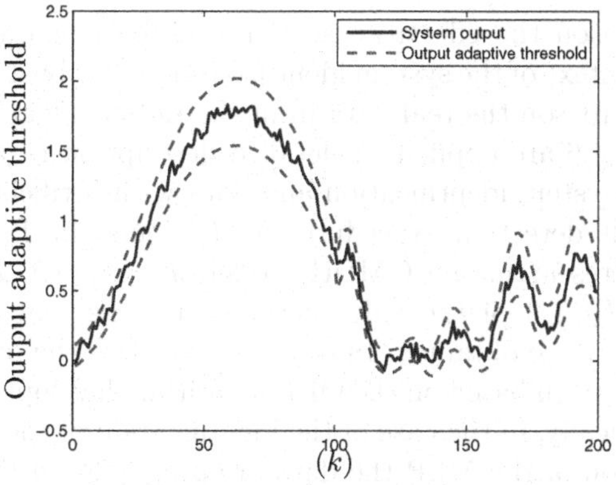

Fig. 5.18. Output adaptive threshold obtained via the OBE

Figure 5.18 presents the output adaptive threshold obtained with the application of the OBE based on the expression (5.31). As it can be seen the adaptive threshold is quite wide what follows from the fact that the ellipsoid approximating the exact feasible parameters set is quite large. Finally, Fig. 5.19 presents the output adaptive threshold calculated with the application of (5.39) for the ZBA. These results show that this approach provides better results than the OBE algorithm.

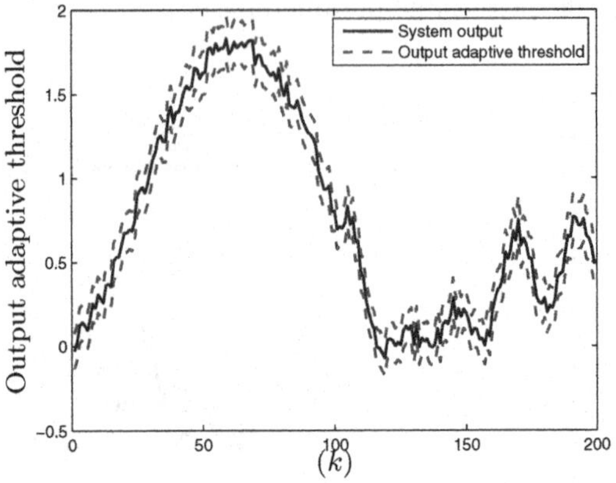

Fig. 5.19. Output adaptive threshold obtained via the ZBA

5.6 GMDH Models in Robust Fault Detection of Valve Actuator

In the present section the effectiveness of the GMDH neural model based approach in the context of the system identification and robust fault detection is shown. For this reason the real data from the Lublin sugar factory widely described in Sect. 4.6 are applied. A detailed description regarding the real data used for the system identification and data with artificially introduced faults for the fault detection procedure can be found in Tab. 4.2. At the first stage of the research, the GMDH models $F = r_F(X, P_1, P_2, T_1)$ and $X = r_X(C_V, P_1, P_2, T_1)$ of the valve actuator for the real data from the Lublin sugar factory are obtained. In order to calculate the output adaptive thresholds the approach based on the OBE algorithm developed in Sect. 3.3.3 is applied. Analogically, in the case of the identification of the valve actuator with the application of the MLP the data are preprocessed by the selection of the every subsequent 10-th sample values resulting in the $n_{\mathcal{T}} = 1000$ training and $n_{\mathcal{V}} = 1000$ validation data sets. Moreover, the output data are transformed taking into account the response range of the hyperbolic tangent activation functions of the neuron output. In order to perform data transformation, a linear scaling is used.

During the GMDH neural network synthesis the dynamic neurons with the IIR filters are used. The neurons are chosen with the application of the SSM. The quality index of a neuron for the validation data set is defined as:

$$Q_{\mathcal{V}} = \frac{1}{n_{\mathcal{V}}} \sum_{k=1}^{n_{\mathcal{V}}} \left| \left(\hat{y}_k^M + \varepsilon_k^M\right) - \left(\hat{y}_k^m + \varepsilon_k^m\right) \right|, \tag{5.47}$$

where $\hat{y}_k^M$ and $\hat{y}_k^m$ are calculated with (5.12) for the first layer or with (5.22)-(5.23) for the subsequent ones. Table 5.1 presents the evolution of (5.47) for the subsequent layers, i.e., these values are obtained for the best performing neurons in a particular layer. Additionally, for the sake of comparison, the results based on the classical quality index (4.41) [118] are presented as well.

Table 5.1. Evolution of $Q_{\mathcal{V}}$ and $B_{\mathcal{V}}$ for the subsequent layers of the GMDH model

Layer	$r_F(\cdot)$ $Q_{\mathcal{V}}$	$r_F(\cdot)$ $B_{\mathcal{V}}$	$r_X(\cdot)$ $Q_{\mathcal{V}}$	$r_X(\cdot)$ $B_{\mathcal{V}}$
1	1.5549	0.3925	0.5198	0.0768
2	1.5277	0.3681	0.4914	0.0757
3	1.5047	0.3514	0.4904	0.0762
4	1.4544	0.3334	0.4898	0.0750
5	1.4599	0.3587	0.4909	0.0748

The results presented in Tab. 5.1 clearly show that the gradual decrease of $Q_{\mathcal{V}}$ occurs when a new layer is introduced. It follows from the fact that the introduction of a new neuron increases the complexity of the model as well as its modelling abilities. On the other hand, if the model is too complex, then the quality index $Q_{\mathcal{V}}$ increases. This situation occurs, for both $F = r_F(\cdot)$ and $X = r_X(\cdot)$, when the 5th layer is introduced. This means that the GMDH neural networks corresponding to $F = r_F(\cdot)$ and $X = r_X(\cdot)$ should have 4 layers. From Tab. 5.1, it can be also seen that the application of the quality index $B_{\mathcal{V}}$ gives similar results for $F = r_F(\cdot)$, i.e., the same number of layers is selected, whilst for $X = r_X(\cdot)$ it leads to the selection of too simple structure, i.e., the neural network with only two layers is selected. It implies that the quality index $Q_{\mathcal{V}}$ makes it possible to obtain a model with smaller uncertainty. In order to achieve the final structure of the juice flow model $F = r_F(\cdot)$ and the servomotor rod displacement $X = r_X(\cdot)$, all unnecessary neurons are removed, leaving only those that are relevant for the computation of the model output.

The final structures of the GMDH neural networks are presented in Figs. 5.20 and 5.21. From Fig. 5.21, it can be seen that the input variable P_2 denoting pressure at the valve outlet, is removed during the model development procedure. Nevertheless, the quality index $Q_{\mathcal{V}}$ achieves a relatively low level. It can be concluded that P_2 has a relatively small influence on the servomotor rod displacement X. This is an example of structural errors that may occur during the selection of neurons in the layer of the GMDH network.

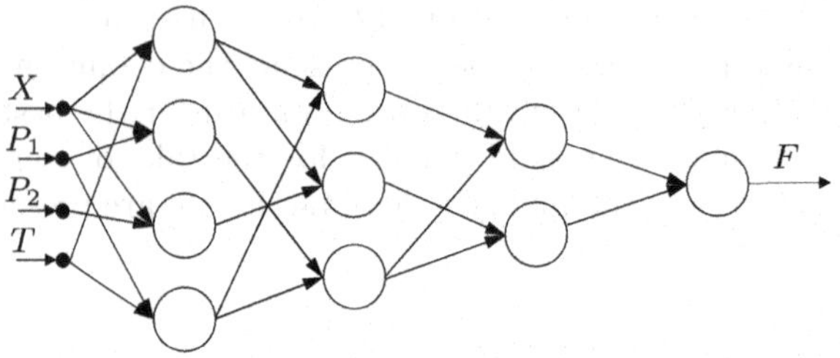

Fig. 5.20. Structure of the juice flow model

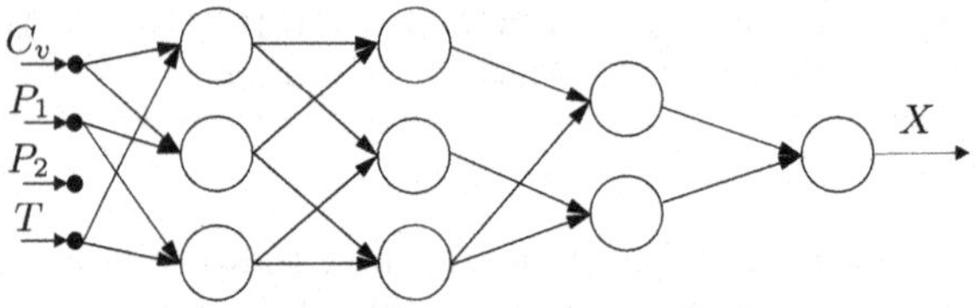

Fig. 5.21. Structure of the servomotor rod displacement model

Figures 5.22 and 5.23 present the modelling abilities of the obtained models $F = r_F(\cdot)$ and $X = r_X(\cdot)$ as well as the corresponding output adaptive threshold obtained with the proposed approach for the validation data set. For reader convenience, Fig. 5.24 presents a selected part of Fig. 5.22 for $k = 400 - 500$ samples. The thick solid line represents the real system output whereas the thin solid lines correspond to the output adaptive threshold, and the dashed line is the model output. From Fig. 5.22 and 5.23, it is clear that the system response is contained within the output adaptive threshold generated according to the proposed approach. It should be pointed out that these system bounds are designed with the estimated output error bounds. The above estimates are $\varepsilon^m_{n_\mathcal{T}} = -0.8631$ and $\varepsilon^M_{n_\mathcal{T}} = 0.5843$ for $F = r_F(\cdot)$ while $\varepsilon^m_{n_\mathcal{T}} = -0.2523$ and $\varepsilon^M_{n_\mathcal{T}} = 0.2331$ for $X = r_X(\cdot)$.

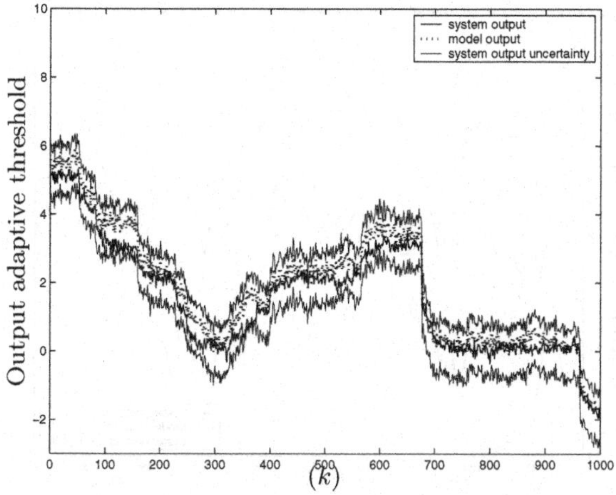

Fig. 5.22. Juice flow and output adaptive threshold

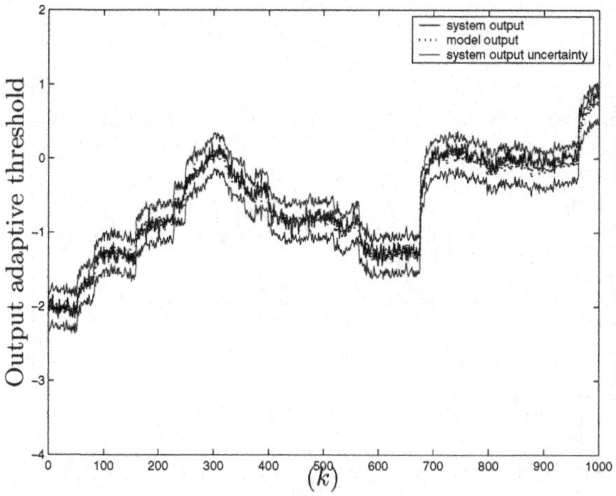

Fig. 5.23. Servomotor rod displacement and output adaptive threshold

As it was mentioned, the quality of the GMDH model can be further improved with the application of the technique described in Sect. 5.4. This technique can be perceived as the retraining method of the network. For the considered valve actuator it is profitable to utilise the retraining technique for the model $F = r_F(\cdot)$. As a result, the quality index (4.41) is decreased from 0.3334 to 0.2160 (cf. Tab. 5.1). These results as well as the comparison of Fig. 5.24 and 5.25 justify the need for the retraining technique proposed in Sect. 5.4.

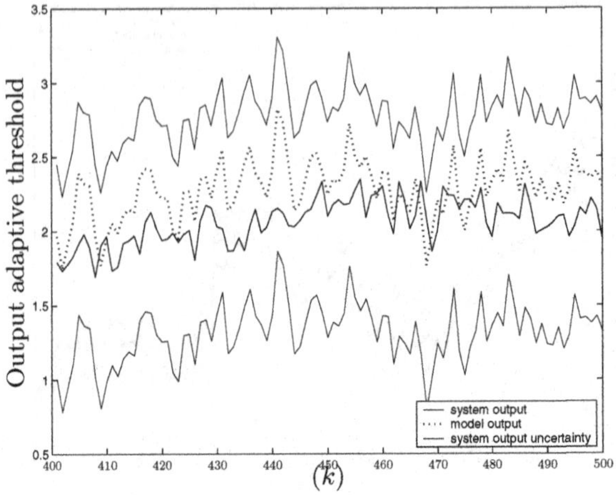

Fig. 5.24. Selected part of the juice flow and output adaptive threshold

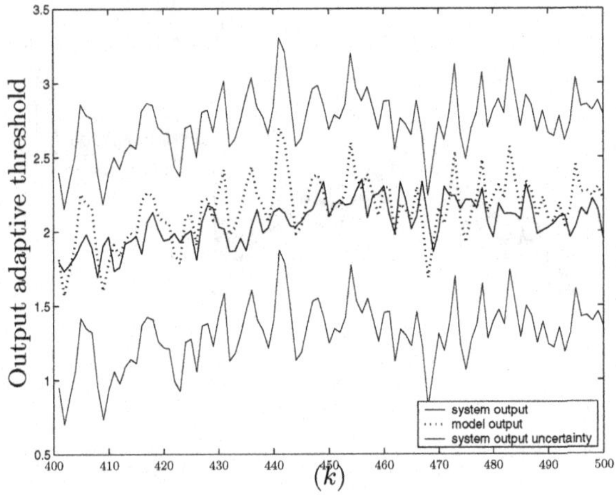

Fig. 5.25. Juice flow and output adaptive threshold after retraining

The main objective of this application study is to develop a fault detection scheme for the considered valve actuator. Since both $F = r_F(\cdot)$ and $X = r_X(\cdot)$ are designed with the approach proposed in Sect. 5.3, it is possible to employ them for the robust fault detection. This task can be realised according to the rules described in Sect. 4.2. Figures 5.26-5.29 present the output adaptive threshold for the faulty data.

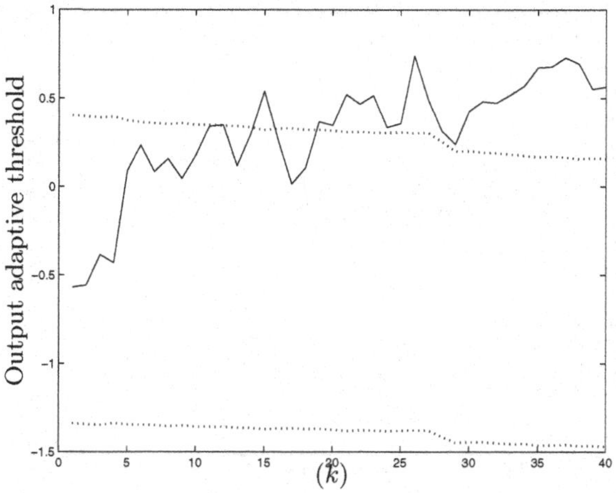

Fig. 5.26. Output adaptive threshold for the fault f_{16}

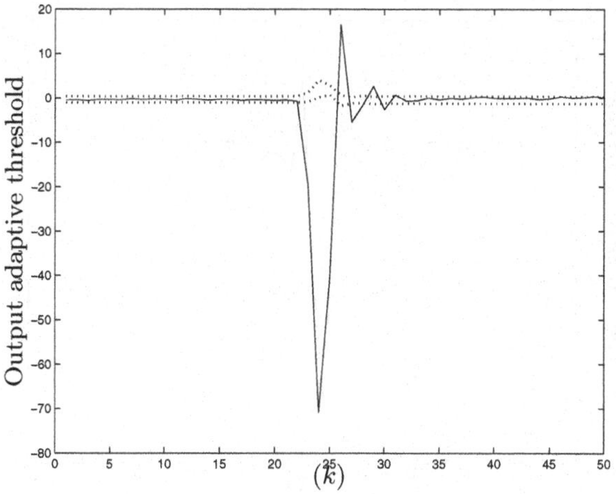

Fig. 5.27. Output adaptive threshold for the fault f_{17}

From these results it can be seen that it is possible to detect all four faults, although fault f_{18} is detected 18 sec. after its occurrence. It is caused by the relative insensitivity of the obtained model to this particular fault. The results presented so far are obtained with the data from a real system. It should be also pointed out that, within the framework of the actuator benchmark [156], the data for only four general faults $f_{16} - f_{19}$ are available.

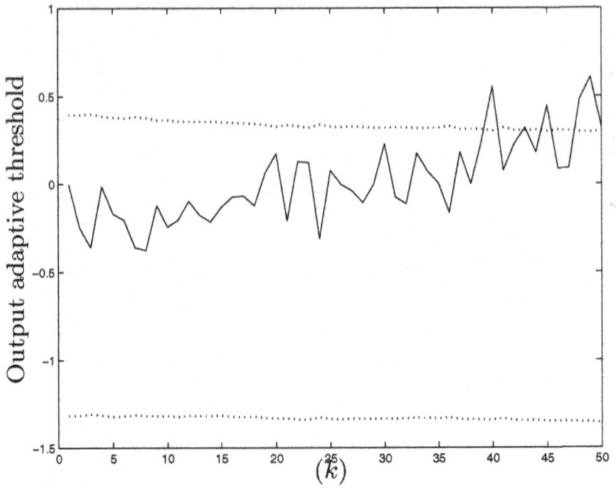

Fig. 5.28. Output adaptive threshold for the fault f_{18}

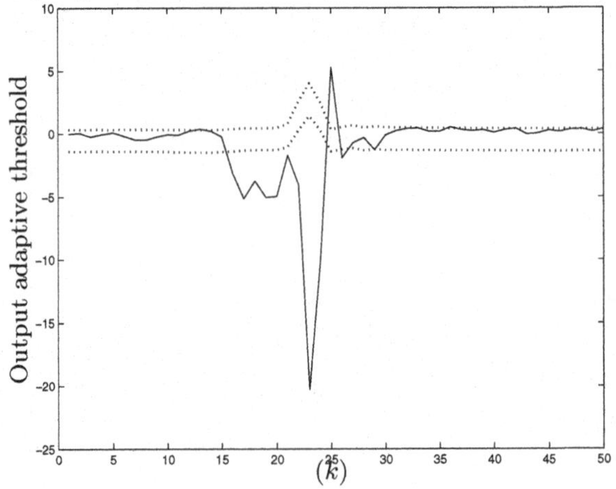

Fig. 5.29. Output adaptive threshold for the fault f_{19}

In order to provide a more comprehensive and detailed application the study of the proposed fault diagnosis scheme, a DAMADICS Matlab Simulink actuator model depicted in Fig 5.30 is employed [156]. This tool makes it possible to generate data for 19 different faults which are described in Tab. 5.3. In the benchmark scenario the abrupt A and incipient I faults are considered. Furthermore, the abrupt faults can be regarded as small S, medium M and big B according to the benchmark description.

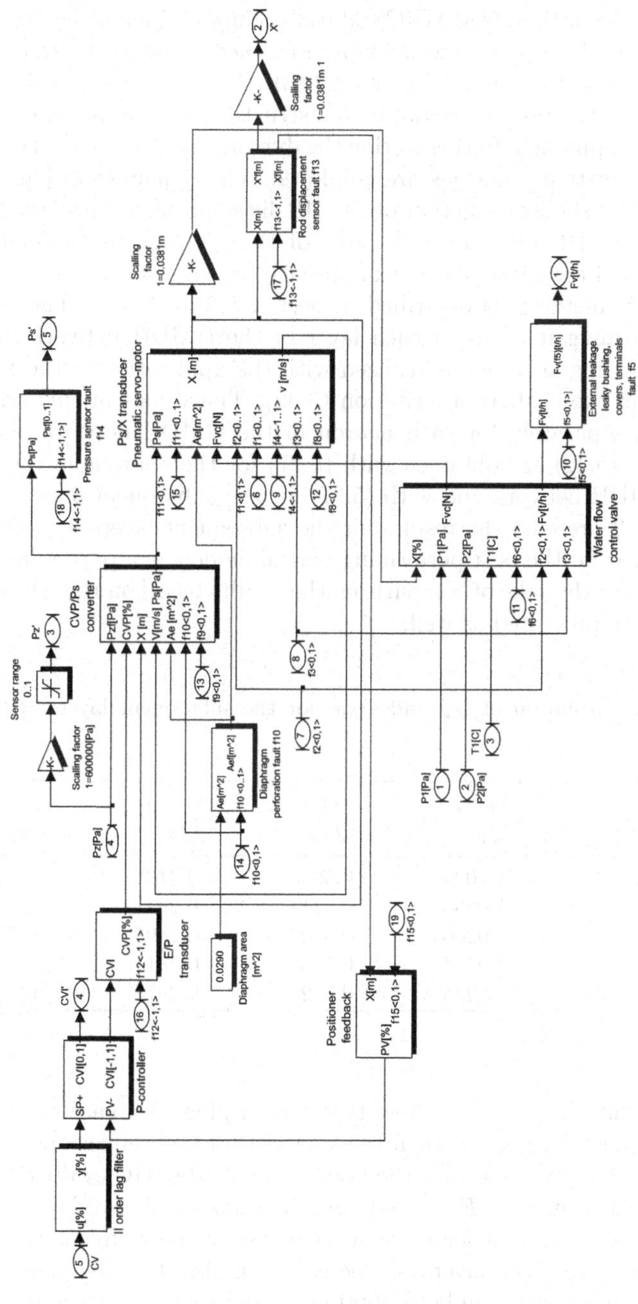

Fig. 5.30. DAMADICS valve actuator implemented in the Matlab Simulink

In the case of the DAMADICS Matlab Simulink model the synthesis process of the GMDH neural network proceeds according to the steps described in Sect. 2.4 with the application of the OBE. As it was mentioned in Sect. 2.4, the choice of the proper partial model structure is an important problem in the GMDH approach. In this section the dynamic neurons with the hyperbolic tangent activation functions are employed. The dynamics in partial models is realized by the introduction of the IIR filter presented in Sect. 2.5.1. The order of the IIR filters in each layer of the network is obtained with the application of the Lipschitz index approach based on so-called Lipschitz quotients. This methods is described in Sect. 2.5.2 in details. The selection of best performing neurons in each layer of the GMDH network in terms of their processing accuracy is realized with the application of the SSM which is based on the evaluation criterion (5.42). The values of this criterion are calculated separately for each neuron in the GMDH network, whereas $\hat{y}_k^m$ and $\hat{y}_k^M$ in (5.42) are obtained with (5.27) for the neurons in the first layer of the GMDH network and with (5.31) for the subsequent ones.

Table 5.2 presents the results for the subsequent layers, i.e., these values are obtained for the best performing partial models in a particular layer. Additionally, for the sake of comparison, the results based on the AIC evaluation criterion are presented as well.

Table 5.2. Evolution of $Q_{\mathcal{V}}$ and Q_{AIC} for the subsequent layers of the GMDH models

Layer	$r_F(\cdot)$ $Q_{\mathcal{V}}$	$r_F(\cdot)$ Q_{AIC}	$r_X(\cdot)$ $Q_{\mathcal{V}}$	$r_X(\cdot)$ Q_{AIC}
1	1.1034	0.5998	0.3198	0.1878
2	1.0633	0.5711	0.2931	0.1756
3	1.0206	0.5525	0.2895	0.1771
4	0.9434	0.5138	0.2811	0.1749
5	1.9938	0.5528	0.2972	0.1719

The results in Tab. 5.2 show that the gradual decrease of value of the evaluation criteria occurs when the synthesis of the network proceed. Moreover, it can be also seen that the application of the AIC evaluation criterion gives similar results for $F = r_F(\cdot)$, i.e., the same number of layers is selected, whilst for $X = r_X(\cdot)$ it leads to the selection of too simple structure, i.e., a network with only two layers is selected. It implies that the quality index $Q_{\mathcal{V}}$ and Q_{AIC} makes it possible to obtain a model with smaller uncertainty.

After the synthesis of the $F = r_F(\cdot)$ and $X = r_X(\cdot)$ models, it is possible to employ them for the robust fault detection. This task can be realised with the application of the output adaptive threshold according to (5.37). Figures 5.31-5.34 present the system responses and the corresponding output adaptive thresholds for the faulty data.

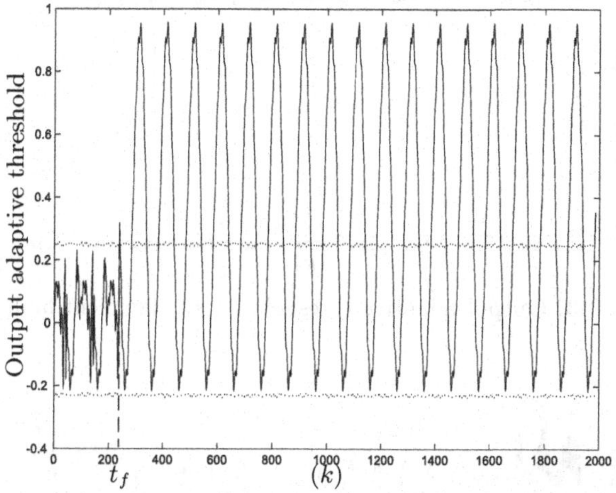

Fig. 5.31. Output adaptive threshold for the big abrupt fault f_1

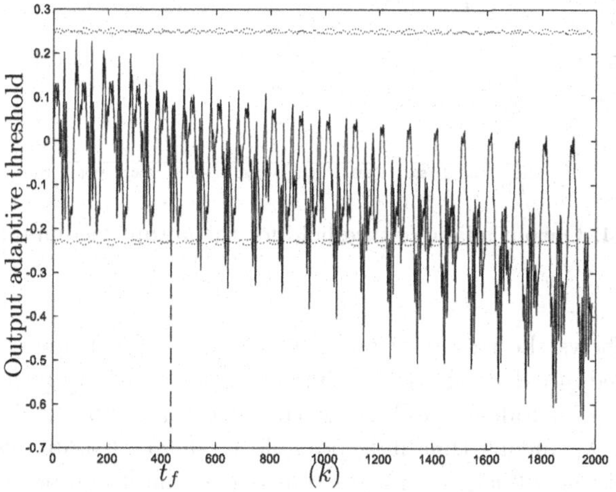

Fig. 5.32. Output adaptive threshold for the incipient fault f_2

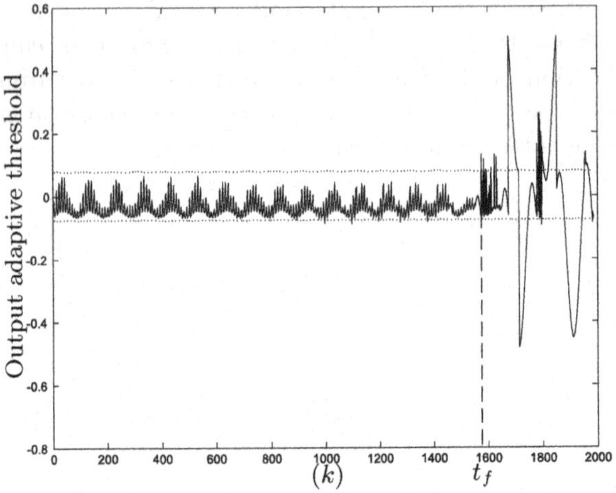

Fig. 5.33. Output adaptive threshold for the incipient fault f_4

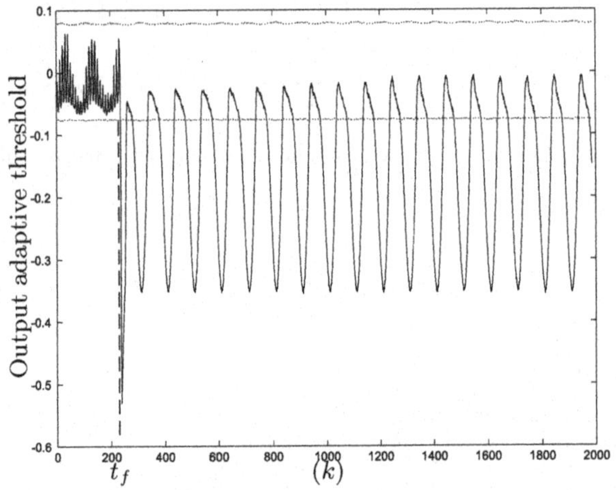

Fig. 5.34. Output adaptive threshold for the abrupt medium fault f_7

Table 5.3 shows the results of fault detection of all considered faults (S – small, M – medium, B – big, I – Incipient). The notation given in Tab. 5.3 can be explained as follows: ND means that it is impossible to detect a given fault, D_F or D_X means that it is possible to detect a fault with $r_F(\cdot)$ or $r_X(\cdot)$, respectively, while D_{FX} means that a given fault can be detected with both $r_F(\cdot)$ or $r_X(\cdot)$. From the results presented in Tab. 5.3 it can be seen that it is impossible to detect faults f_5, f_9, and f_{14}. Moreover, some small and medium faults cannot be detected too, i.e., f_8 and f_{12}. This situation can be explained by the fact that the effect of these faults is at the same level as the effect of noise.

Table 5.3. Results of robust fault detection of the valve actuator

f	Faults description	S	M	B	I
f_1	Valve clogging	ND	$D_{F,X}$	$D_{F,X}$	
f_2	Valve plug or valve seat sedimentation			D_F	D_F
f_3	Valve plug or valve seat erosion				D_F
f_4	Increased of valve or busing friction				D_X
f_5	External leakage				ND
f_6	Internal leakage (valve tightness)				D_F
f_7	Medium evaporation or critical flow	$D_{F,X}$	D_X	D_X	
f_8	Twisted servomotor's piston rod	ND	ND	D_X	
f_9	Servomotors housing or terminals tightness				ND
f_{10}	Servomotor's diaphragm perforation	D_X	D_{FX}	D_{FX}	
f_{11}	Servomotor's spring fault			D_X	D_{FX}
f_{12}	Electro-pneumatic transducer fault	ND	ND	D_X	
f_{13}	Rod displacement sensor fault	D_F	D_F	D_F	D_{FX}
f_{14}	Pressure sensor fault	ND	ND	ND	
f_{15}	Positioner feedback fault			D_X	
f_{16}	Positioner supply pressure drop	D_F	D_X	D_X	
f_{17}	Unexpected pressure change across the valve			D_{FX}	D_{FX}
f_{18}	Fully or partly opened bypass valves	D_F	D_F	D_F	D_F
f_{19}	Flow rate sensor fault	D_F	D_F	D_F	

5.7 Concluding Remarks

The present section proposes a complete design procedure concerning the application of the GMDH neural networks to the robust fault detection. Starting from a set of input-output measurements of the system, it is shown how to estimate the parameters and the corresponding uncertainty of a neuron via the LMS, BEA, OBE and ZBA algorithm. The methodology developed for the parameter and uncertainty estimation of a neuron makes it possible to formulate an algorithm that allows to obtain a neural networks with a relatively small modelling uncertainty. All the hard computation regarding the design of the GMDH neural network are performed off-line and hence, the problem regarding the time consuming calculations is not of paramount importance.

Based on the GMDH neural network, a novel robust fault detection scheme is proposed which supports the diagnostic decisions. The proposed approach for system identification and fault detection is tested on the DAMADICS benchmark problem [156]. The experimental results show that the approach makes it possible to obtain a suitably accurate mathematical description of the system with the small modelling uncertainty. The developed models are employed to design the fault detection scheme for the valve actuator. As a result of the fault detection study with the real data from the actuator, it can be said that it is possible to detect all faults being considered, i.e., a group of faults. In order to perform a more comprehensive study, the Matlab Simulink simulator of the valve actuator is employed. The obtained results show that almost all faults can be detected except for a few incipient or small ones. It indicates that a technique that enables a further increase in the fault sensitivity should be developed. Undoubtedly, such a technique should decrease modelling uncertainty. For example, it can be realised by the application of the experimental design approach [21].

6

State-Space GMDH Networks for Actuator Robust FDI

6.1 Introduction

The description of the neural model in the suitable state-space representation is one of the most desirable feature in the context of the application of such a model in the FDI systems [1, 2, 3, 4, 5, 6, 7, 8, 14]. Indeed, the dynamic non-linear neural model obtained during the system identification can be applied for the robust FDI by the application of the classical filters which require a state-space system description. The state-space representation of the dynamic neuron presented in Sect. 2.5.4, together with the fact that in the GMDH model the parameters of each neuron are estimated separately allows easily to apply advanced parameters estimation algorithms, e.g., the UKF [166, 167]. The training method, which is based on the UKF constrained parameters estimation, warranties the asymptotically stable neurons and as a consequence the stability of the whole GMDH network as well. Another advantage of the UKF is that it allows to obtain a description of the neural model uncertainty, which then can be applied to the robust fault detection with the application of the output adaptive thresholds.

The approach developed in this chapter allows to perform the robust fault detection, however, it does not enable to distinguish which system component causes the fault. In order to perform the fault isolation the approach relying on the application of the bank of neural models, which generates the residuals for each fault, is usually used. In practice such an approach is difficult to realize because it is often not possible to provide the data allowing to design the neural models of particular faults. In order to partly solve such a challenging problem, new approaches of the robust FDI of the actuators are presented. The developed methods are based on the estimation of the GMDH neural model inputs with the application of the UIF [14, 22, 60, 168] and RUIF [62]. These methods allow to obtain the adaptive thresholds for

M. Mrugalski, *Advanced Neural Network-Based Computational Schemes for Robust Fault Diagnosis*, Studies in Computational Intelligence 510,
DOI: 10.1007/978-3-319-01547-7_6,

each input signal of the diagnosed system. In the consequence this approach enables to perform the robust FDI of the actuators in a non-linear systems simultaneously [61, 62].

Especially, the RUIF-based method is attractive from the point of view of its application to the actuators robust FDI. The developed estimation methodology can be perceived as a combination of linear-system strategies [169] and [170] for a class of non-linear systems [171, 172]. It should be also pointed out that the proposed description of non-linearity constitutes an alternative to the approaches presented in the literature [173, 174, 175]. The main contribution is the proposed design procedure of an observer-based FDI scheme for which a prescribed disturbance attenuation level is achieved with respect to the input estimation error while guaranteeing the convergence of the observer with a possibly large decay rate of the state estimation error. However, the main advantage of the proposed approach boils down to its simplicity. Indeed, it is shown that it reduces to solve a set of linear matrix inequalities, which can be efficiently solved with computational packages, e.g., with the Matlab.

The chapter is organized as follows: Section 6.2 contains an original concept of the robust FDI of actuators. Section 6.3 presents novel developments regarding the synthesis process of the state-space GMDH neural model. Section 6.4 shows the details of the application of the UKF to parameters and uncertainty estimation of the asymptotically stable state-space GMDH neural model. Moreover, the original method of calculation of the output adaptive thresholds for the system and sensors robust fault detection is developed. Sections 6.5 and 6.6 present innovative developments regarding calculation of the input adaptive thresholds with the application of the UIF and RUIF. In Sect. 6.7 an illustrative example shows the effectiveness of the UKF and RUIF in the confidence estimation of the neuron outputs and inputs. Section 6.8 provides an example of the application of the proposed approach to the robust FDI of a tunnel furnace. It should be also underlined pointed out that the presented results are based on [61, 62, 77, 176].

6.2 Robust Fault Detection and Isolation of Actuators

The diagnosed system often consists of a process and some amount of actuators and sensors (cf. Fig. 6.1). The developed in Chaps. 4 and 5 methods allows to perform the robust fault detection on the basis of the output adaptive thresholds according to the scheme presented in Fig. 6.2. These outputs adaptive thresholds should contain real system responses in fault-free mode. An occurrence of the system or sensor faults are signaled when system outputs $\boldsymbol{y}_k$ crosses the output adaptive threshold (cf. Fig. 6.3):

$$\hat{y}_{i,k}^{m} \leq y_{i,k} \leq \hat{y}_{i,k}^{M}, \tag{6.1}$$

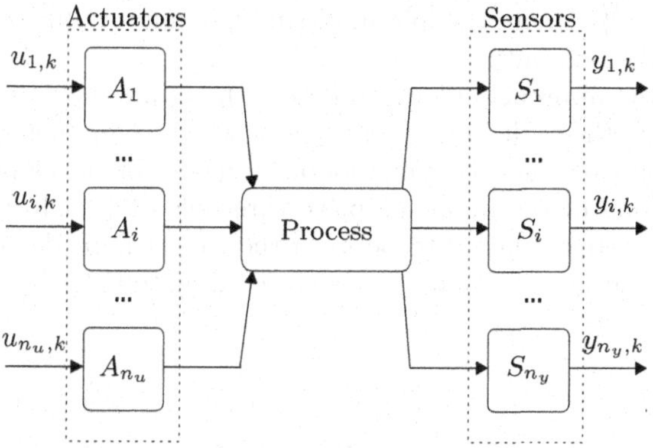

Fig. 6.1. Scheme of the system with multiple actuators and sensors

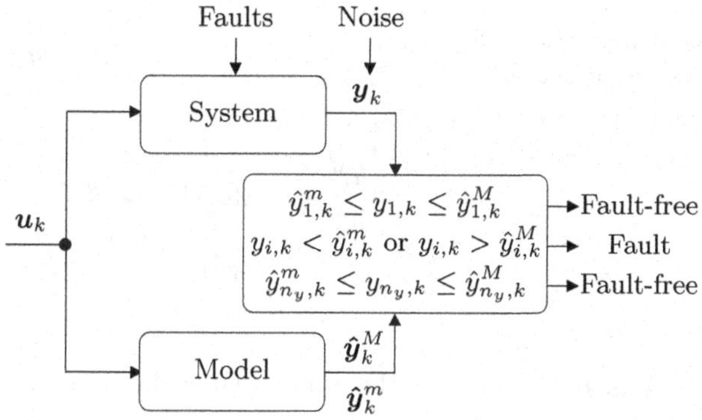

Fig. 6.2. Scheme of the robust fault detection of the system and sensors

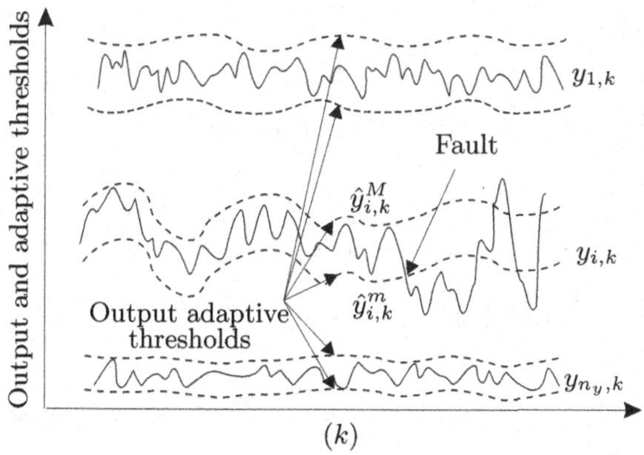

Fig. 6.3. Output adaptive thresholds in the robust fault detection of the system and sensors

where $\hat{y}^m_{i,k}$ and $\hat{y}^M_{i,k}$ denote the minimum and maximum value of the adaptive threshold for the i-th output.

The presented approach allows to detect the faults, however, it does not allow to perform the fault isolation and show which component of the system is faulty. For this reason a new methodology has to be developed. Following it, the calculation of the input adaptive thresholds for each input signal of the diagnosed system enables to perform the robust fault detection and the isolation of the actuators simultaneously according to the scheme depicted in Fig. 6.4.

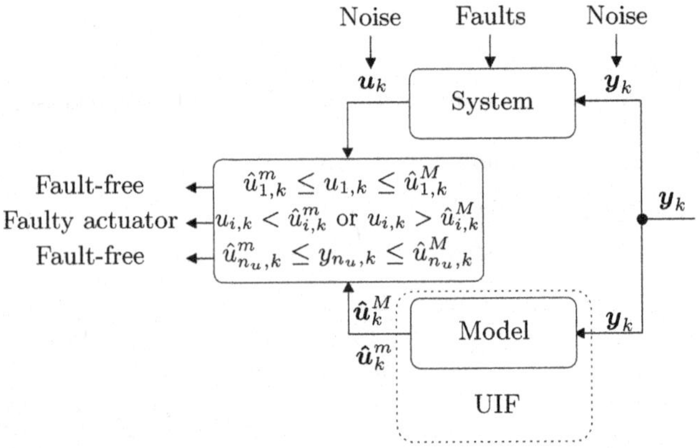

Fig. 6.4. Scheme of the robust FDI of the actuators

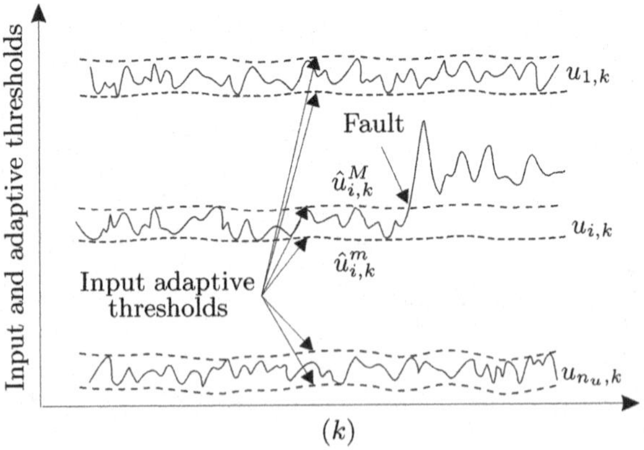

Fig. 6.5. Input adaptive thresholds in the robust FDI of actuators

In order to achieve this goal the methodology of the calculation of the inputs adaptive thresholds has to be developed (cf. Fig. 6.5):

$$\hat{u}_{i,k}^{m} \leq u_{i,k} \leq \hat{u}_{i,k}^{M}, \tag{6.2}$$

where $\hat{u}_{i,k}^{m}$ and $\hat{u}_{i,k}^{M}$ represent the minimum and maximum value of the adaptive threshold for the i-th system input. To achieve this goal, the UIF can be used, however, the description of the neural model in the state-space representation is required.

6.3 Synthesis of State-Space GMDH Neural Network

Let us assume that in the general case each neuron in the GMDH network has the following form:

$$\hat{\boldsymbol{s}}_{i,j,k}^{(l)} = \mathcal{F}\left(\boldsymbol{r}_{i,k}^{(l)}, \boldsymbol{p}_{i,j}^{(l)}\right), \tag{6.3}$$

where: $\boldsymbol{r}_{i,k}^{(l)} \in \mathbb{R}^{n_r}$ for $i = 1, ..., n_R$ are the neuron input vectors formed as the combinations of the neural model inputs $\boldsymbol{r}_{i,k}^{(l)} = [u_{i,k}^{(l)}, \ldots, u_{j,k}^{(l)}]^T$, $\hat{\boldsymbol{s}}_{i,j,k}^{(l)} \in \mathbb{R}^{n_s}$ for $j = 1, ..., n_N$ are the neurons outputs vectors formed as the combinations of the network outputs $[\hat{y}_{i,k}^{(l)}, \ldots, \hat{y}_{j,k}^{(l)}]^T$, $\boldsymbol{p}_{i,j}^{(l)} \in \mathbb{R}^{n_r \times n_s}$ denotes the parameters estimate matrix, $\mathcal{F}(\cdot)$ is the activation function, and l is the number of layers in the GMDH network.

The process of the synthesis of the first layer of the GMDH neural network begins from the creation of a set of n_R vectors of the neuron inputs $\boldsymbol{r}_{i,k}^{(l)}$ based on the combinations of the model inputs $u_k \in \mathbb{R}^{n_u}$ belonging to the training data set $\mathcal{T}$. The number of the vectors $\boldsymbol{r}_{i,k}^{(l)}$ depends on the number of the model inputs n_u and the number of the neuron inputs n_r. The structure of the layer of the GMDH network is presented in Fig. 6.6. Each i-th vector $\boldsymbol{r}_{i,k}^{(l)}$ constitutes the neurons stimulation which results in the formation of j-th neurons and their outputs $\hat{\boldsymbol{s}}_{i,j,k}^{(l)}$, which are the estimates of the modeled system outputs. The number n_N of these neurons, for the each subsequent i-th vector $\boldsymbol{r}_{i,k}^{(l)}$, depends on the number of modeled output signals n_y and the assumed number of the neurons inputs n_r:

$$\begin{cases} \hat{\boldsymbol{s}}_{1,1,k}^{(1)} & = \mathcal{F}(\boldsymbol{r}_{1,k}^{(1)}) \\ & \cdots \\ \hat{\boldsymbol{s}}_{1,n_N,k}^{(1)} & = \mathcal{F}(\boldsymbol{r}_{1,k}^{(1)}) \\ & \cdots \\ \hat{\boldsymbol{s}}_{n_R,1,k}^{(1)} & = \mathcal{F}(\boldsymbol{r}_{n_R,k}^{(1)}) \\ & \cdots \\ \hat{\boldsymbol{s}}_{n_R,n_N,k}^{(1)} & = \mathcal{F}(\boldsymbol{r}_{n_R,k}^{(1)}). \end{cases} \tag{6.4}$$

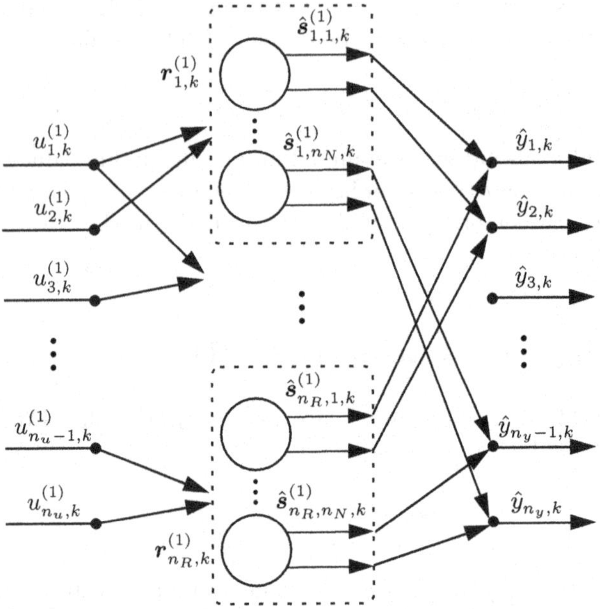

Fig. 6.6. First layer of the state-space GMDH model

In order to estimate an unknown parameters of the dynamic neurons the UKF [167] can be applied. In section 6.4, it is shown that the UKF-based constrained parameters estimation warranties the asymptotically stable neurons of the GMDH model. Moreover, the application of this algorithm to the parameters estimation process enables to obtain the uncertainty of the partial models, simultaneously. After the estimation, the parameters of the neurons are not modified during the further network synthesis. The obtained parameters estimates and their uncertainty enable to calculate the neuron responses and the adaptive threshold which can be applied in the robust fault detection scheme.

At the next stage of the GMDH network synthesis, a validation data set $\mathcal{V}$ is used to calculate the processing error $Q(\hat{\mathbf{s}}^{(l)}_{i,j})$ of each partial model in the current l-th network layer:

$$Q = \begin{bmatrix} Q(\hat{\mathbf{s}}^{(l)}_{1,1,k}) & \dots & Q(\hat{\mathbf{s}}^{(l)}_{1,j,k}) & \dots & Q(\hat{\mathbf{s}}^{(l)}_{1,n_N,k}) \\ \dots & \dots & \dots & \dots & \dots \\ Q(\hat{\mathbf{s}}^{(l)}_{i,1,k}) & \dots & Q(\hat{\mathbf{s}}^{(l)}_{i,j,k}) & \dots & Q(\hat{\mathbf{s}}^{(l)}_{i,n_N,k}) \\ \dots & \dots & \dots & \dots & \dots \\ Q(\hat{\mathbf{s}}^{(l)}_{n_R,1,k}) & \dots & Q(\hat{\mathbf{s}}^{(l)}_{n_R,j,k}) & \dots & Q(\hat{\mathbf{s}}^{(l)}_{n_R,n_N,k}) \end{bmatrix}. \tag{6.5}$$

Based on the chosen evaluation criteria [118], it is possible to select the best-fitted neurons in the layer (cf. Fig. 6.7). The selection method allows to optimise the structure of a new layer of neurons and prevents to uncontrolled increase of the GMDH neural network structure.

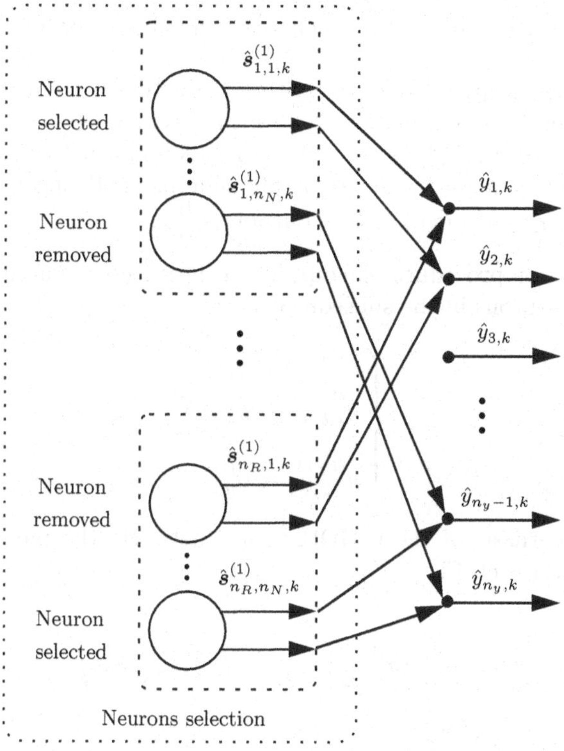

Fig. 6.7. Selection in the state-space GMDH model

According to the chosen selection method elements, which introduce too big processing error, are removed. In order to achieve this goal the following modified method based on the SSM presented in Sect. 2.4.1 can be applied:

Input : Q – the matrix of the quality indexes of all dynamic neurons in the l-th layer, n_o – the number of opponent neurons, n_w – the number of winnings required for i-th neuron selection.
Output : The set of neurons after the selection.

1. Select $j = 1$ column of the matrix Q representing the quality indexes of all n_R neurons modeling j-th vector of the system outputs $\boldsymbol{s}_{i,j,k}$ created on the basis of all $i = 1, \ldots, n_R$ vectors of the system inputs $\boldsymbol{r}_{i,k}$.

2. Conduct series of n_y competitions between each i-th neuron in the j-th column and n_o randomly selected neurons so-called opponent from the same column. The i-th neuron is so-called winner neuron when:

$$Q(\hat{\boldsymbol{s}}_{i,1,k}^{(l)}) \leq Q_o(\hat{\boldsymbol{s}}_{i,1,k}^{(l)}), \tag{6.6}$$

where $o = 1, \ldots, n_o$ and Q_o denotea a quality index of the opponent neuron;
3. Select the neurons for the $(l+1)$-th layer with the number of winnings bigger than n_w (the remaining neurons are removed);
4. Repeat the steps 1–3 for $j = 2, \ldots, n_N$ column of the matrix Q representing the quality indexes of all neurons modeling the remaining $j = 2, \ldots, n_N$ vectors of system outputs $\hat{\boldsymbol{s}}_{i,1,k}^{(l)}$.

After the selection procedure, the outputs of the selected neurons become the inputs of the neurons in the subsequent layer:

$$\begin{cases} u_{1,k}^{(l+1)} &= \hat{y}_{1,k}^{(l)} \\ u_{2,k}^{(l+1)} &= \hat{y}_{2,k}^{(l)} \\ & \ldots \\ u_{n_u,k}^{(l+1)} &= \hat{y}_{n_y,k}^{(l)}. \end{cases} \tag{6.7}$$

During the synthesis of the GMDH neural network, the number of layers suitably increases (cf. Fig. 6.8).

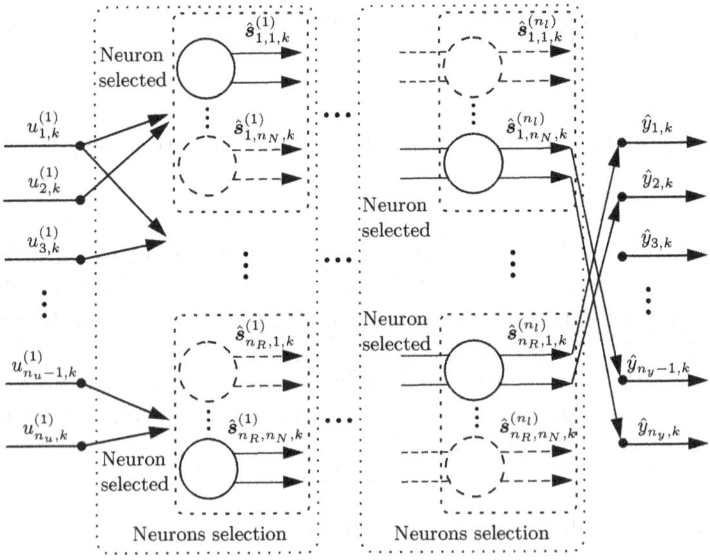

Fig. 6.8. Synthesis of the state-space GMDH model

At the next stage of the GMDH network synthesis the termination condition testing is performed. For this reason the quality indexes $Q(\hat{\boldsymbol{s}}_{i,j}^{(l)})$ for all neurons included in the l layer are calculated. The quality index $Q_{j,min}^{(l)}$ represents the processing error for the best neuron in this layer which approximates the j-th vector of system outputs:

$$Q_{j,min}^{(l)} = \min_{i=1,\dots,n_R} Q(\hat{\boldsymbol{s}}_{i,j}^{(l)}) \quad \text{for} \quad j = 1, \dots, n_N. \tag{6.8}$$

The values $Q(\hat{\boldsymbol{s}}_{i,j}^{(l)})$ can be determined with the application of the defined evaluation criterion which is used in the selection process. The synthesis of the network is completed when each of the calculated quality indexes reaches the minimum:

$$Q_{j,min}^{(l_{opt})} = \min_{l=1,\dots,n_l} Q_{j,min}^{(l)} \quad \text{for} \quad j = 1, \dots, n_N. \tag{6.9}$$

The termination of the synthesis appears independently for each vector of the system outputs $\hat{\boldsymbol{s}}_{i,j}^{(l)}$ and as a result a set of quality indexes, corresponding to each vector of the system outputs is obtained $Q_1, Q_2, \dots, Q_{n_N}$. The particular minimum can occur at different stages of the network synthesis. It results from that in the multi-output network, the outputs of the resulting structure are usually in different layers (cf. Fig. 6.9).

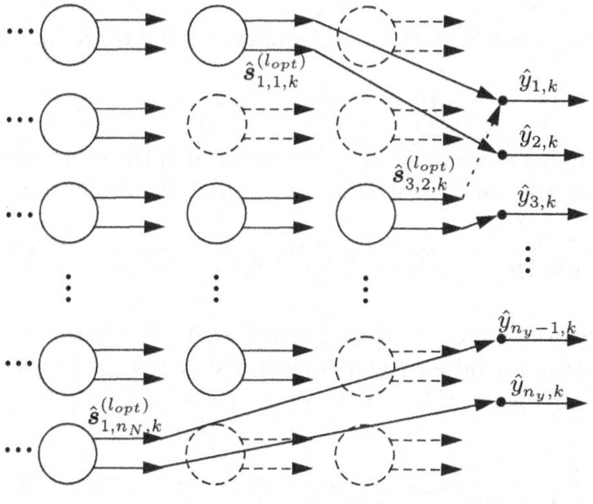

Fig. 6.9. Final structure of the state-space GMDH model

As a result of the process of the GMDH network synthesis the set of vectors of potential system outputs is obtained. It may happen that in these vectors there are a few outputs modeling the particular system output. In this case, the neuron output, which characterizes the best modeling quality, should be chosen.

6.4 UKF in Confidence Estimation of GMDH Model Outputs

In Sect. 2.4 it was mentioned that the parameters of each neuron in the GMDH neural network are estimated separately. This property allows to apply the UKF in the process of the synthesis of the GMDH neural network. It is shown that the main advantage of application of the UKF to the parameters estimation of the dynamic neurons with the constraints warranties the asymptotically stable GMDH model. Moreover, the application of the UKF allows to calculate the model uncertainty, which then can be applied to the robust fault detection. Let us define a state vector (in order to simplify the notation, the indexes $_{i}^{(l)}$ in the $\boldsymbol{r}_{i,k}^{(l)}$ and $_{i,j}^{(l)}$ in $\hat{\boldsymbol{s}}_{i,j,k}^{(l)}$ are omitted):

$$\boldsymbol{x}_k = \left[\boldsymbol{p}_k \boldsymbol{z}_k\right]^T, \tag{6.10}$$

which is composed of the parameter vector of the neuron $\boldsymbol{p}_k$ as well as of the state of the neuron, which is described in a form:

$$\boldsymbol{z}_{k+1} = \boldsymbol{A}(\boldsymbol{p}_k)\boldsymbol{z}_k + \boldsymbol{B}(\boldsymbol{p}_k)\boldsymbol{r}_k, \tag{6.11}$$

$$\tilde{\boldsymbol{s}}_k = \boldsymbol{C}(\boldsymbol{p}_k)\boldsymbol{z}_k, \tag{6.12}$$

$$\hat{\boldsymbol{s}}_k = \mathcal{F}(\tilde{\boldsymbol{s}}_k). \tag{6.13}$$

The vector $\boldsymbol{p}_k$ is composed of the diagonal elements of the matrix $\boldsymbol{A}$, i.e.,

$$\boldsymbol{p}_k = \left[a_{11}, ..., a_{n,n}, ...\right]^T, \tag{6.14}$$

while the remaining elements of $\boldsymbol{p}_k$ are composed of the remaining parameters of $\boldsymbol{A}$, as well as all elements of $\boldsymbol{B}$ and $\boldsymbol{C}$. Thus, the dimension of $\boldsymbol{p}_k$ is:

$$\dim(\boldsymbol{p}_k) = \frac{(n_z \times n_z) + n_z}{2} + n_z \times n_r + n_s = n_p. \tag{6.15}$$

It should be also pointed out that instead of $\boldsymbol{A}$ $(\boldsymbol{B}, \boldsymbol{C})$ the notation $\boldsymbol{A}(\boldsymbol{p}_k)$ $(\boldsymbol{B}(\boldsymbol{p}_k), \boldsymbol{C}(\boldsymbol{p}_k))$ is introduced which clearly denotes the dependence on $\boldsymbol{p}_k$. Finally, the state-space model of the dynamic neuron is:

$$\boldsymbol{x}_{k+1} = \begin{bmatrix} \boldsymbol{p}_k \\ \boldsymbol{A}(\boldsymbol{p}_k)\boldsymbol{z}_k + \boldsymbol{B}(\boldsymbol{p}_k)\boldsymbol{r}_{i,k}^{(l)} \end{bmatrix} + \boldsymbol{\varpi}_k = \mathcal{G}(\boldsymbol{x}_k, \boldsymbol{r}_{i,k}^{(l)}) + \boldsymbol{\varpi}_k, \tag{6.16}$$

$$\hat{\boldsymbol{s}}_{i,j,k}^{(l)} = \mathcal{F}(\boldsymbol{C}(\boldsymbol{p}_k)\boldsymbol{z}_k) + \boldsymbol{\upsilon}_k = \mathcal{H}(\boldsymbol{x}_k) + \boldsymbol{\upsilon}_k. \tag{6.17}$$

where $\mathcal{G} : \mathbb{R}^n \times \mathbb{R}^{n_r} \to \mathbb{R}^n$ and $\mathcal{H} : \mathbb{R}^n \to \mathbb{R}^{n_s}$ are the process and observation models, respectively. $\boldsymbol{\varpi}_{k-1} \in \mathbb{R}^n$ is the process noise, and $\boldsymbol{\upsilon}_0 \in \mathbb{R}^n$ is the measurement noise. It is assumed that the process noise and the measurement noise are uncorrelated. $\rho(\boldsymbol{x}_0)$, $\rho(\boldsymbol{\varpi}_{k-1})$, $\rho(\boldsymbol{\upsilon}_k)$ are the Probability Density Function (PDF), where $\boldsymbol{x}_0 \in \mathbb{R}^n$ is the initial state vector. Moreover, mean and covariance of $\rho(\boldsymbol{\varpi}_{k-1})$ and $\rho(\boldsymbol{v}_k)$ are known and equal to zero and $\boldsymbol{Q}$, $\boldsymbol{R}$, respectively.

The profit function, which is the value of the conditional PDF of the state vector $\boldsymbol{x}_k \in \mathbb{R}^n$ given the past and present measured data $\boldsymbol{s}_1, \ldots, \boldsymbol{s}_k$, is defined as follows:

$$J(\boldsymbol{x}_k) \triangleq \rho(\boldsymbol{x}_k | (\boldsymbol{s}_1, \ldots, \boldsymbol{s}_k)). \tag{6.18}$$

The parameter and state estimation problem can be defined as the maximization of (6.18). In order to solve the following problem the UKF can be applied. The UKF employs the unscented transform [177], which approximates the mean $\hat{\boldsymbol{s}}_k \in \mathbb{R}^{n_s}$ and covariance $\boldsymbol{P}_k^{ss} \in \mathbb{R}^{n_s \times n_s}$ of so-called transformed sigma points after the non-linear transformation $\boldsymbol{s}_k = \mathcal{H}(\boldsymbol{x}_k)$, where the mean and covariance of sigma points are given as $\hat{\boldsymbol{x}}_k \in \mathbb{R}^n$ and $\boldsymbol{P}_k^{xx} \in \mathbb{R}^{n \times n}$. The scheme of the UKF algorithm is presented in Fig. 6.10.

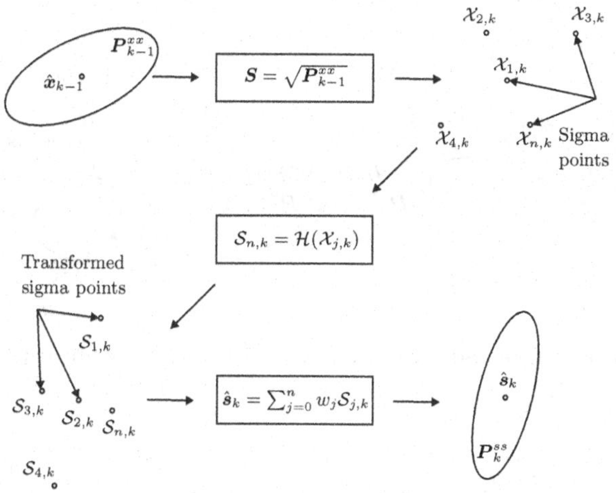

Fig. 6.10. Scheme of the UKF algorithm

The UKF [178] can be perceived a derivative-free alternative to the Extended Kalman Filter (EKF) in the framework of the state-estimation. The UKF calculates the mean and covariance of a random variable, which undergoes a non-linear transformation by utilizing a deterministic "sampling" approach. Generally, $2n + 1$, sigma points are chosen on the basis of a square-root decomposition of the prior covariance. These sigma points are propagated through the nonlinearity, without any approximation, and then a weighted mean (estimate of neuron output) and covariance can be calculated. In Figs. 6.11-6.13 the examples of covariance and mean propagation through real nonlinearity, the EKF and UKF are presented. It can be noticed that in the case of the UKF there is almost no bias error in the estimate of the mean and the estimated covariance is also much closer to the real covariance. Opposite to the UKF, in the case of the EKF the errors in both the mean and covariance are calculated by a first-order approximation. It follows from the fact that, it is easier to approximate a probability distribution than an arbitrary non-linear function.

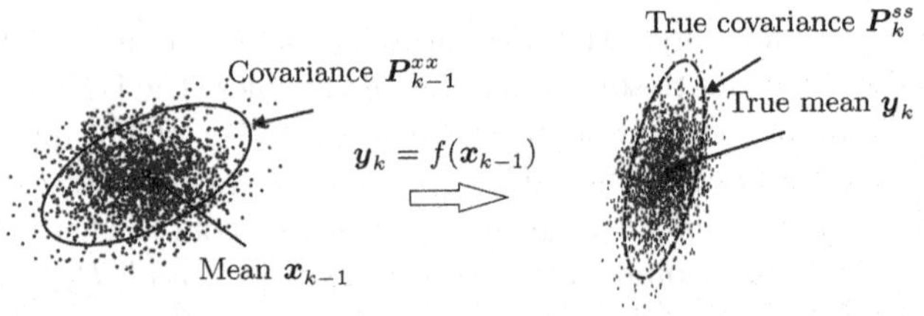

Fig. 6.11. Covariance and mean propagation through real nonlinearity

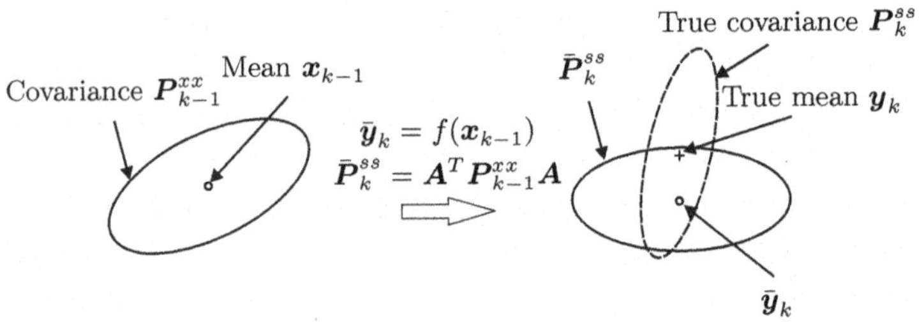

Fig. 6.12. Covariance and mean propagation through nonlinearity via the EKF

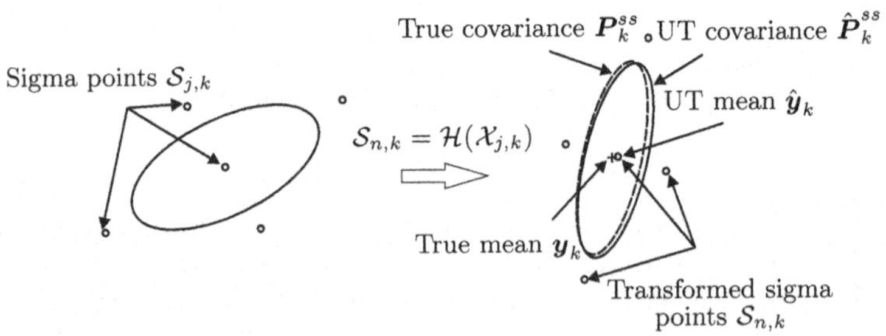

Fig. 6.13. Covariance and mean propagation through nonlinearity via the UKF

The UKF involves a recursive application of sigma points to the state-space equations. The standard UKF implementation for state-estimation uses the following variable definitions: $w_0^{n_s} = \lambda/(n+\lambda)$, $w_0^{c} = \lambda/(n+\lambda)+(1-\alpha^2+\beta)$, $w_j^{n_s} = w_j^{c} = 1/\{2(n+\lambda)\}$, $\lambda = L(\alpha^2-1)$ and $\eta = \sqrt{(n+\lambda)}$, where w_j $(j=1,\ldots,2n)$ is a set of scalar weights, λ and η are scaling parameters. The constant α determines the spread of sigma points around $\hat{x}$ and is usually set to $1e-4 \leq \alpha \leq 1$. β is used to incorporate a prior knowledge of the distribution (for Gaussian distribution $\beta = 2$ is an optimal choice [177]). The UKF algorithm is as follows:

1. Initialize with:

$$\hat{\boldsymbol{x}}_0 = \mathcal{E}[\boldsymbol{x}_0] \quad \boldsymbol{P}_0^{xx} = \mathcal{E}[(\boldsymbol{x}_0 - \hat{\boldsymbol{x}}_0)(\boldsymbol{x}_0 - \hat{\boldsymbol{x}}_0)^T], \tag{6.19}$$

for $k \in \{1, \ldots, \infty\}$ calculate $2n+1$ sigma points:

$$\begin{aligned}\hat{\mathcal{X}}_{k-1} =& [\hat{\boldsymbol{x}}_{k-1}, \hat{\boldsymbol{x}}_{k-1} + \eta S(1), \ldots, \hat{\boldsymbol{x}}_{k-1} + \eta S(n), \\ & \hat{\boldsymbol{x}}_{k-1} - \eta S(1), \ldots, \hat{\boldsymbol{x}}_{k-1} - \eta S(n)],\end{aligned} \tag{6.20}$$

where $\boldsymbol{S} = \sqrt{\boldsymbol{P}_{k-1}^{xx}}$ and $\boldsymbol{S}(j)$ stands for the j-th column of $\boldsymbol{S}$.

2. Forecast step:

$$\hat{\mathcal{X}}_{j,k|k-1} = \mathcal{G}(\hat{\mathcal{X}}_{j,k-1|k-1}, \boldsymbol{r}_{k-1,k-1}), \quad j = 0, \ldots, 2n, \tag{6.21}$$

where $\mathcal{G}$ is defined by the right side of (6.16).

$$\hat{\boldsymbol{x}}_{k,k-1} = \sum_{j=0}^{2n} w_j^{n_s} \hat{\mathcal{X}}_{j,k|k-1}, \tag{6.22}$$

$$\boldsymbol{P}_{k,k-1}^{xx} = \sum_{j=0}^{2n} w_j^c [\hat{\mathcal{X}}_{j,k|k-1} - \hat{\boldsymbol{x}}_{k,k-1}][\hat{\mathcal{X}}_{j,k|k-1} - \hat{\boldsymbol{x}}_{k,k-1}]^T + \boldsymbol{Q}, \tag{6.23}$$

$$\hat{\mathcal{S}}_{j,k|k-1} = \mathcal{H}(\hat{\mathcal{X}}_{j,k|k-1}), \quad j = 0, \ldots, 2n, \tag{6.24}$$

$$\hat{\boldsymbol{s}}_{k|k-1} = \sum_{j=0}^{2n} w_j^{n_s} \hat{\mathcal{S}}_{j,k|k-1}, \tag{6.25}$$

where $\boldsymbol{P}_{k|k-1}^{xx}$ is the forecast error-covariance.

3. Data assimilation step:

$$\begin{aligned}\boldsymbol{P}_k^{ss} &= \boldsymbol{C}\boldsymbol{P}_{k,k-1}^{xx}\boldsymbol{C}^T + \boldsymbol{R}, && (6.26)\\ \boldsymbol{K}_k &= \boldsymbol{P}_{k,k-1}^{xx}\boldsymbol{C}^T(\boldsymbol{P}_k^{ss})^{-1}, && (6.27)\\ \hat{\boldsymbol{s}}_{k|k-1} &= \boldsymbol{C}\hat{\boldsymbol{x}}_{k|k-1}, && (6.28)\\ \hat{\boldsymbol{x}}_k &= \hat{\boldsymbol{x}}_{k|k-1} + \boldsymbol{K}_k(\boldsymbol{s}_k - \hat{\boldsymbol{s}}_{k|k-1}) && (6.29)\\ \boldsymbol{P}_k^{xx} &= \left[\boldsymbol{I}_n - \boldsymbol{P}_{k,k-1}^{xx}\boldsymbol{K}_k\boldsymbol{C}\right]\boldsymbol{P}_{k,k-1}^{xx}. && (6.30)\end{aligned}$$

where $\boldsymbol{P}_{k|k}^{ss}$ is the data assimilation error-covariance and $\boldsymbol{K}_k \in \mathbb{R}^{n \times n_s}$ is the Kalman gain matrix.

The task of training of dynamic neuron relies on the estimation of parameters vector $\boldsymbol{x}_k$ which satisfies the following interval constraint:

$$-1+\delta \leq \boldsymbol{e}_i^T \boldsymbol{x}_k \leq 1-\delta, \quad i=1,...,n \tag{6.31}$$

where: $\boldsymbol{e}_i \in \mathbb{R}^{n_p+n}$ whereas $\boldsymbol{e}_1 = [1,0,...,0]^T$, $\boldsymbol{e}_2 = [0,1,...,0]^T$, ect., and δ is a small positive value. These constrains follow directly from the asymptotic stability condition (2.86), while δ is introduced in order to make the above mentioned problem tractable.

The neural model has a cascade structure what follows from the fact that the neuron outputs constitute the neuron inputs in the subsequent layers. The neural model, which is the result of the cascade connection of dynamic neurons is asymptotically stable, when each of neurons is asymptotically stable [127]. So, a fulfilment of (2.86) (being a result of (6.31)) for each neuron allows to obtain an asymptotically stable dynamic GMDH neural model. Thus, the objective of the interval-constrained parameter-estimation problem is to maximize (6.18) subject to (6.31).

In order to perform the neuron training process it is necessary to truncate the PDF at the n constraint edges given by the rows of the state interval constraint (6.31) such that the pseudomean $\hat{\boldsymbol{x}}_{k,k}^t$ of the truncated PDF is an interval-constrained state estimate with the truncated error covariance $\boldsymbol{P}_{k,k}^{xx}$ (cf. Fig 6.14). The procedure of the truncation of the PDF allows to avoid the explicit on-line solution of a constrained optimisation problem at each time step. Moreover, it assimilates the interval-constraint information

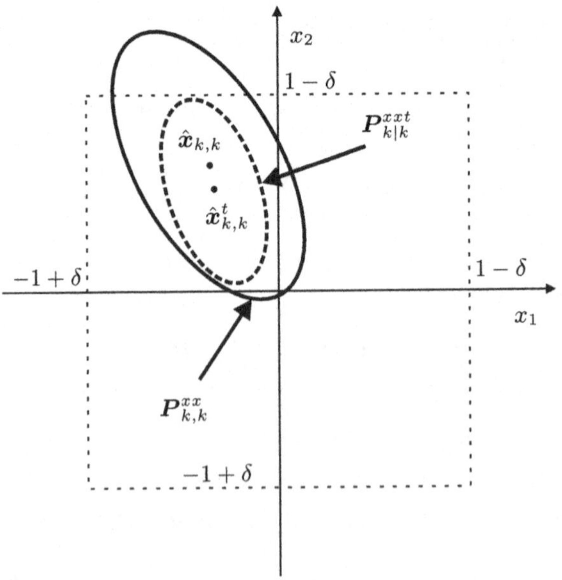

Fig. 6.14. Truncation of the PDF

in the state estimate $\hat{\boldsymbol{x}}^{t}_{k|k}$ and error covariance $\boldsymbol{P}^{xxt}_{k|k}$. The procedure of PDF truncation [167] can be performed in i steps, where $i = 1, \ldots, n$. Let $\hat{\boldsymbol{x}}^{t}_{k|k,i}$ and $\boldsymbol{P}^{xxt}_{k|k,i}$ denote, respectively, the state estimate and the error covariance after enforcing the first $i = 1$ rows of the state interval constraint (6.31). On the beginning, at $i = 1$ the initialization of $\hat{\boldsymbol{x}}^{t}_{k|k,i} = \hat{\boldsymbol{x}}_{k|k}$ and $\boldsymbol{P}^{xxt}_{k|k,i} = \boldsymbol{P}^{xx}_{k|k}$ according to (6.30) is performed. Next, for $i = 1, \ldots, n$, perform the following procedure:

1. Find the orthogonal matrix $\boldsymbol{S} \in \mathbb{R}^{n \times n}$ and diagonal matrix $\boldsymbol{D} \in \mathbb{R}^{n \times n}$ from the Schur decomposition [179] of $\boldsymbol{P}^{xxt}_{k|k,i}$ given by $\boldsymbol{S}\boldsymbol{D}\boldsymbol{S}^{T} = \boldsymbol{P}^{xxt}_{k|k}$ where $\boldsymbol{P}^{xxt}_{k|k,i}$ is symmetric.
2. Perform Gram-Schmidt orthogonalization to find the orthogonal matrix $\boldsymbol{\Theta} \in \mathbb{R}^{n \times n}$ satisfying:

$$\boldsymbol{\Theta}\boldsymbol{D}^{1/2}\mathrm{col}_i(\boldsymbol{S}^T) = \left[\sqrt{\boldsymbol{P}^{xxt}_{(i,i),k|k,i}}\ \boldsymbol{0}_{1\times(n-1)}\right]^T, \tag{6.32}$$

where for $l = 1$ is given by:

$$\mathrm{row}_l(\boldsymbol{\Theta}) = \frac{1}{\sqrt{\boldsymbol{P}^{xxt}_{(i,i),k|k,i}}}\mathrm{row}_i(\boldsymbol{S})\boldsymbol{D}^{1/2}, \tag{6.33}$$

and for $l = 2, ..., n$ is given by:

$$\mathrm{row}_l(\boldsymbol{\Theta}) = (e_l - \sum_{q=1}^{l-1}(e_l^T\mathrm{col}_q(\boldsymbol{\Theta}^T))\mathrm{col}_q(\boldsymbol{\Theta}^T))^T, \tag{6.34}$$

where $e_l \triangleq \mathrm{col}_l(\boldsymbol{I}_{n\times n})$, if $\mathrm{row}_l(\boldsymbol{\Theta}) = \boldsymbol{0}_{1\times n}$ then reset

$$\mathrm{row}_l(\boldsymbol{\Theta}) = (e_1 - \sum_{q=1}^{l-1}(e_1^T\mathrm{col}_q(\boldsymbol{\Theta}^T))\mathrm{col}_q(\boldsymbol{\Theta}^T))^T, \tag{6.35}$$

also, normalize for $l = 1, ..., n$

$$\mathrm{row}_l(\boldsymbol{\Theta}) = \frac{1}{\|\mathrm{row}_l(\boldsymbol{\Theta})\|_2}\mathrm{row}_l(\boldsymbol{\Theta}). \tag{6.36}$$

3. Find the parameters of the interval constraint $a_{k,i} \leq \varsigma_{i,k,i} \leq b_{k,i}$, where $a_{k,i} < b_{k,i} \in \mathbb{R}$ are given by the following expressions:

$$a_{k,i} = \frac{1}{\sqrt{\boldsymbol{P}^{xxt}_{(i,i),k|k,i}}}(d_{i,k} - x^t_{i,k|k,i}), \tag{6.37}$$

$$b_{k,i} = \frac{1}{\sqrt{\boldsymbol{P}^{xxt}_{(i,i),k|k,i}}}(e_{i,k} - x^t_{i,k|k,i}), \tag{6.38}$$

and $\varsigma_{i,k} \triangleq \boldsymbol{\Theta}\boldsymbol{D}^{-1/2}\boldsymbol{S}^T(\boldsymbol{x}_k - \hat{\boldsymbol{x}}^t_{k|k,i}) \in \mathbb{R}^n$ with the mean:

$$\hat{\varsigma}_{k,i} = [\mu_i \;\; \boldsymbol{0}_{1\times(n-1)}]^T, \tag{6.39}$$

and the covariance:

$$\boldsymbol{P}^{\varsigma\varsigma}_{k,i} = \text{diag}(\sigma_i^2, \boldsymbol{1}_{1\times(n-1)}), \tag{6.40}$$

where:

$$\alpha_i = \frac{\sqrt{2}}{\sqrt{\pi}[\text{erf}(\frac{b_{k,i}}{\sqrt{2}}) - \text{erf}(\frac{a_{k,i}}{\sqrt{2}})]}, \tag{6.41}$$

whereas erf$(\cdot)$ is the error function defined as:

$$\text{erf}(t) \triangleq \frac{2}{\sqrt{\pi}}\int_0^t \exp(-\tau^2)d\tau, \tag{6.42}$$

and

$$\mu_i = \alpha_i[\exp(\frac{-a^2_{k,i}}{2}) - \exp(\frac{-b^2_{k,i}}{2})], \tag{6.43}$$

$$\sigma_i^2 = \alpha_i[\exp(\frac{-a^2_{k,i}}{2})(a_{k,i}-2\mu_i) - \exp(\frac{-b^2_{k,i}}{2})(b_{k,i}-2\mu_i))] + \mu_i^2 + 1. \tag{6.44}$$

4. Perform the inverse transformation

$$\hat{\boldsymbol{x}}^t_{k|k,i+1} = \boldsymbol{S}\boldsymbol{D}^{1/2}\boldsymbol{\Theta}^T\hat{\varsigma}_{k,i} + \hat{\boldsymbol{x}}^t_{k|k,i}, \tag{6.45}$$

$$\boldsymbol{P}^{xx,t}_{k|k,i+1} = \boldsymbol{S}\boldsymbol{D}^{1/2}\boldsymbol{\Theta}^T\boldsymbol{P}^{\varsigma\varsigma}_{k,i}\boldsymbol{\Theta}\boldsymbol{D}^{1/2}\boldsymbol{S}^T. \tag{6.46}$$

and write i instead n obtaining $\hat{\boldsymbol{x}}^t_{k|k} = \hat{\boldsymbol{x}}^t_{k|k,n+1}$ and $\boldsymbol{P}^{xxt}_{k|k} = \boldsymbol{P}^{xxt}_{k|k,n+1}$.

The application of the UKF with the procedure of the truncation of the PDF allows to obtain the state estimates as well as the uncertainty of the GMDH model in the form of matrixes $\boldsymbol{P}^{xxt}$ which can be then applied to the calculation of the system outputs adaptive thresholds and perform the robust fault detection according to the scheme presented in Fig. 6.2. The real system responses in the fault-free mode should be contained in the following output adaptive thresholds:

$$\hat{y}^m_{i,k} \leq y_{i,k} \leq \hat{y}^M_{i,k}, \tag{6.47}$$

where $\hat{y}^m_{i,k}$ and $\hat{y}^M_{i,k}$ are calculated according to following equations:

$$\hat{y}^m_{i,k} = \mathcal{F}_i\left(\boldsymbol{c}_i\hat{\boldsymbol{x}}_k - t^{\alpha/2}_{n_t-n_p}\hat{\sigma}_i\sqrt{\boldsymbol{c}_i\boldsymbol{P}^{xxt}\boldsymbol{c}_i^T}\right), \tag{6.48}$$

and

$$\hat{y}^M_{i,k} = \mathcal{F}_i\left(\boldsymbol{c}_i\hat{\boldsymbol{x}}_k + t^{\alpha/2}_{n_t-n_p}\hat{\sigma}_i\sqrt{\boldsymbol{c}_i\boldsymbol{P}^{xxt}\boldsymbol{c}_i^T}\right), \tag{6.49}$$

where $\boldsymbol{c}_i$ stands for the i-th row ($i = 1, ..., n_s$) of the matrix $\boldsymbol{C}$ of the output neuron, $\hat{\sigma}_i$ is the standard deviation of the i-th fault-free residual and $t^{\alpha/2}_{n_t-n_p}$ is the t-Student distribution quantile. An occurrence of the system or sensor faults is signaled when the system outputs $\boldsymbol{y}_k$ crosses the output adaptive threshold defined by (6.47) what is presented in Fig. 6.15.

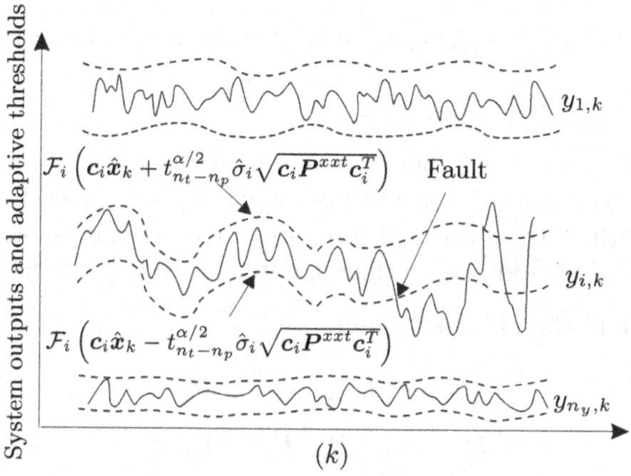

Fig. 6.15. Output adaptive thresholds obtained via the UKF

6.5 UIF in Confidence Estimation of GMDH Model Inputs

The methodology presented in Subsects. 6.3 and 6.4 allows to obtain the scheme of the robust fault detection of the system and sensors with the application of the output adaptive thresholds. It is obtained by the development of the process of the synthesis of the GMDH neural network in the state-space representation with the application of the UKF. This result is important because the state-space representation of the each neurons in the GMDH neural network allows to apply the UIF in order to estimate the input signals of the each neurons and in the consequence the whole GMDH neural network [61]. In the same time it is possible to calculate the input adaptive thresholds for the each input of the diagnosed system. This technique enables to perform the robust FDI of the systems actuators according to the scheme described in the subsection 6.2.

The presented in Sect. 2.5.4 the neuron model can be shown as the canonical observer representation [180]:

$$\boldsymbol{x}_{k+1} = \boldsymbol{A}\boldsymbol{x}_k + \boldsymbol{B}\boldsymbol{r}_k + \boldsymbol{\omega}_k, \tag{6.50}$$

$$\hat{\boldsymbol{s}}_{k+1} = \mathcal{F}(\boldsymbol{C}\boldsymbol{x}_{k+1}) + \boldsymbol{v}_{k+1}, \tag{6.51}$$

where: $\boldsymbol{x}_k \in \mathbb{R}^n$ denotes the system state, $\boldsymbol{r}_k \in \mathbb{R}^{n_r}$ and $\hat{\boldsymbol{s}}_k \in \mathbb{R}^{n_s}$ represent the neuron input and output, $\boldsymbol{A} \in \mathbb{R}^{n_z \times n_z}$, $\boldsymbol{B} \in \mathbb{R}^{n_z \times n_r}$ and $\boldsymbol{C} \in \mathbb{R}^{n_s \times n_z}$ are the known matrices, $\boldsymbol{\omega}_k$ and υ_{k+1} are the process and measurement noise.

Because $\mathcal{F}(\cdot)$ (i.e., $f_i(\cdot) = \tanh(\cdot)$, $i = 1, \ldots, n_s$ are selected) is invertible, let us assume that:

$$\tilde{\boldsymbol{s}}_k = \mathcal{F}^{-1}(\hat{\boldsymbol{s}}_k), \tag{6.52}$$

therefore, (6.51) can be replaced by (6.52), which is linear with respect to the state:

$$\tilde{\boldsymbol{s}}_{k+1} = \boldsymbol{C}\boldsymbol{x}_{k+1} + \upsilon_{k+1}. \tag{6.53}$$

To estimate the state and input, the results of the work [60] are used. The recursive filter is proposed and the filter equations are given below. The recursive part of the filter consists of three steps: the estimation of the unknown input, measurement update and time update. These three steps are given by:

1. Estimation of the unknown input:

$$\tilde{\boldsymbol{R}}_k = \boldsymbol{C}\boldsymbol{P}^{xx}_{k|k-1}\boldsymbol{C}^T + \boldsymbol{R}_k, \tag{6.54}$$

$$\boldsymbol{H}_k = (\boldsymbol{F}_k^T \tilde{\boldsymbol{R}}_k^{-1} \boldsymbol{F}_k)^{-1} \boldsymbol{F}_k^T \tilde{\boldsymbol{R}}_k^{-1}, \tag{6.55}$$

$$\hat{\boldsymbol{r}}_k = \boldsymbol{H}_k(\tilde{\boldsymbol{s}}_k - \boldsymbol{C}\hat{\boldsymbol{x}}_{k|k-1}), \tag{6.56}$$

   where: $\boldsymbol{F}_k = \boldsymbol{C}\boldsymbol{B}$.

2. Prediction:

$$\hat{\boldsymbol{x}}_{k+1|k} = \boldsymbol{A}_k\hat{\boldsymbol{x}}_{k|k} + \boldsymbol{B}\hat{\boldsymbol{r}}_k, \tag{6.57}$$

$$\begin{aligned}\boldsymbol{P}^{xx}_{k|k-1} =& (\boldsymbol{I}_n - \boldsymbol{B}\boldsymbol{H}_k\boldsymbol{C})\boldsymbol{P}^{xx}_{k-1|k-1}(\boldsymbol{I}_n - \boldsymbol{B}\boldsymbol{H}_k\boldsymbol{C})^T + \\ & \boldsymbol{B}\boldsymbol{H}_k\boldsymbol{R}_k\boldsymbol{H}_k^T\boldsymbol{B}^T.\end{aligned} \tag{6.58}$$

3. Update:

$$\boldsymbol{K}_k = (\boldsymbol{P}^{xx}_{k|k-1}\boldsymbol{C}^T - \boldsymbol{B}\boldsymbol{H}_k\boldsymbol{R}_k)\alpha_k^T(\alpha_k\tilde{\boldsymbol{R}}_{k|k-1}\alpha_k^T)^{-1}\alpha_k, \tag{6.59}$$

$$\tilde{\boldsymbol{R}}_{k|k-1} = (\boldsymbol{I}_p - \boldsymbol{C}\boldsymbol{B}\boldsymbol{H}_k)\tilde{\boldsymbol{R}}_k(\boldsymbol{I}_p - \boldsymbol{C}\boldsymbol{B}\boldsymbol{H}_k)^T, \tag{6.60}$$

$$\hat{\boldsymbol{x}}_{k|k-1} = \hat{\boldsymbol{x}}_{k|k-1}\boldsymbol{B}\boldsymbol{s}, \tag{6.61}$$

$$\hat{\boldsymbol{x}}_{k|k} = \hat{\boldsymbol{x}}_{k|k-1} + \boldsymbol{K}_k(\tilde{\boldsymbol{s}}_k - \boldsymbol{C}\hat{\boldsymbol{x}}_{k|k-1}), \tag{6.62}$$

$$\boldsymbol{P}^{xx}_{k|k} = \boldsymbol{P}^{xx}_{k|k-1} - \boldsymbol{K}_k(\boldsymbol{P}^{xx}_{k|k-1}\boldsymbol{C}_k^T + \boldsymbol{S}_{k|k-1})^T, \tag{6.63}$$

$$\alpha_k = [\boldsymbol{0}\boldsymbol{I}_r]\boldsymbol{U}_k^T\tilde{\boldsymbol{S}}_k^{-1} \in \mathbb{R}^{r \times n_s}. \tag{6.64}$$

where: $\hat{\boldsymbol{x}} \in \mathbb{R}^n$ denotes the state vector estimate, $\boldsymbol{P} \in \mathbb{R}^{n \times n}$, $\boldsymbol{R} \in \mathbb{R}^{n_s \times n_s}$ and $\boldsymbol{Q} \in \mathbb{R}^{n \times n}$ are the state vector estimate, measurement noise and process noise covariance matrixes, $\boldsymbol{K} \in \mathbb{R}^{n \times n}$ is the gain matrix. Furthermore, $\boldsymbol{I}_n \in \mathbb{R}^{n \times n}$, $\boldsymbol{I}_{n_y} \in \mathbb{R}^{n_y \times n_y}$, $\text{rank}(\boldsymbol{C}\boldsymbol{B}) = \text{rank}(\boldsymbol{B}) = n_r$ and $\text{rank}(\tilde{\boldsymbol{R}})_{k|k-1} = r$. Moreover, $\boldsymbol{U}_k$ is the orthogonal matrix containing the left singular vectors of

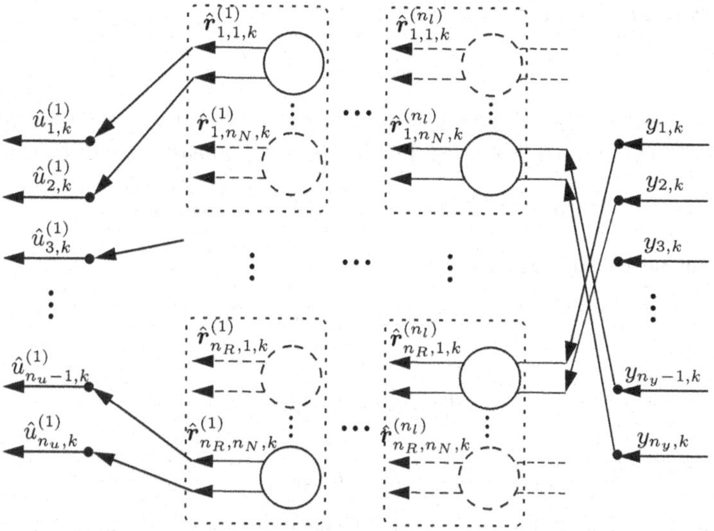

Fig. 6.16. Estimation of the system inputs via the GMDH model and UIF

$\tilde{\boldsymbol{S}}_k^{-1}\boldsymbol{C}_k\boldsymbol{B}_{k-1}$ in its columns and $\tilde{\boldsymbol{S}}_k \in \mathbb{R}^{n_s \times n_s}$ is the invertible matrix satisfying $\tilde{\boldsymbol{S}}_k\tilde{\boldsymbol{S}}_k^T = \tilde{\boldsymbol{R}}_k$, which can be always found by the Cholesky factorization. Note that the time and measurement update of the state estimate take the form of the UIF, except that the true value of the input is replaced by an optimal estimate.

The equations (6.54)-(6.64) enable the estimation of the input signals of the neurons in the last layer of the neural network. Moreover, these signals represent the outputs of the neurons from the previous layer of the network (cf. Fig. 6.16). It should be underlined that the estimated signals have to be contained in the bounded interval. It follows from the fact that the responses of the neurons from the previous layer are limited by the non-linear activation function of the neuron e.g., in the case of the tangent activation function this interval is bounded by the values $a = -1$ and $b = 1$. Thus, the task of training of the dynamic neuron relies on the estimation of neuron inputs which satisfies the following interval constraint:

$$\boldsymbol{a}_k \leq \boldsymbol{r}_k \leq \boldsymbol{b}_k, \tag{6.65}$$

where $\boldsymbol{a}_k \in \mathbb{R}^n$ and $\boldsymbol{b}_k \in \mathbb{R}^n$, $a_{j,k} \leq b_{j,k}$ for $j = 1, \ldots, n_r$ are the known minimum and maximum of allowed values of the signal generated by the neurons from the previous layer of the network. This signal can be defined as follows:

$$\boldsymbol{r}_k = \boldsymbol{H}_k\tilde{\boldsymbol{s}}_{k+1} - \boldsymbol{Z}\boldsymbol{x}_k, \tag{6.66}$$

where: $\boldsymbol{Z} = \boldsymbol{H}_k \boldsymbol{C} \boldsymbol{A}$. Let us assume that:

$$r_{i,k} = h_i^T \tilde{s}_{k+1} - z_i^T x_k, \tag{6.67}$$

where: h_i^T and z_i^T denote i-th rows of matrixes $\boldsymbol{H}$ and $\boldsymbol{Z}$, respectively. Substituting (6.67) into (6.65), the following interval constraint can be obtained:

$$a_k + h_i^T \tilde{s}_{k+1} \leq z_i^T x_k \leq b_k + h_i^T \tilde{s}_{k+1}. \tag{6.68}$$

which can be written as follows:

$$d_k \leq z_i^T x_k \leq e_k, \tag{6.69}$$

where:

$$d_k = a_k + h_i^T \tilde{s}_{k+1}, \tag{6.70}$$

and:

$$e_k = b_k + h_i^T \tilde{s}_{k+1}. \tag{6.71}$$

In order to perform the estimation of the neurons inputs and its adaptive thresholds once again it is necessary to apply the procedure of the PDF truncation [167] at the n constraint edges given by the rows of the state interval constraint (6.69) where $\hat{\boldsymbol{x}}_{k,k}$ and $\boldsymbol{P}_{k,k}^{xx}$ are described by (6.62) and (6.63) respectively, the pseudo-mean and pseudo-covariance are obtained from the UIF.

The application of the UIF allows to obtain estimates of the GMDH neural networks inputs and knowing that the covariance matrix of the input estimate is $\boldsymbol{P} = (\boldsymbol{F}^T \tilde{\boldsymbol{R}}_k^{-1} \boldsymbol{F})^{-1}$ the adaptive thresholds for the inputs of the GMDH neural model receive the following form:

$$\hat{u}_{i,k}^m \leq u_{i,k} \leq \hat{u}_{i,k}^M, \tag{6.72}$$

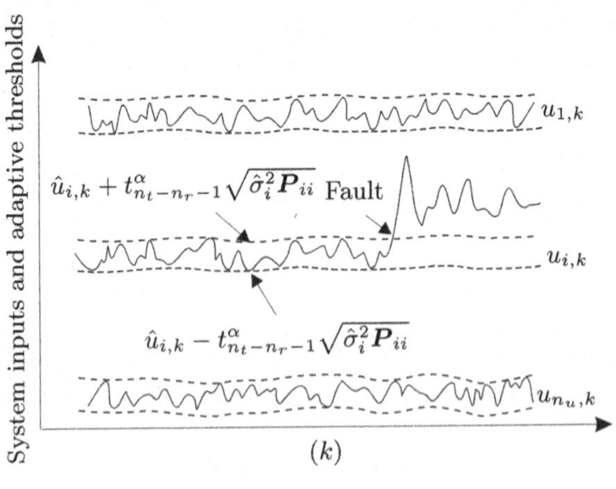

Fig. 6.17. Input adaptive thresholds obtained via the UIF

where:

$$\hat{u}_{i,k}^{m} = \hat{u}_{i,k} - t_{n_t-n_r-1}^{\alpha}\sqrt{\hat{\sigma}_i^2 \boldsymbol{P}_{ii}}, \tag{6.73}$$

and

$$\hat{u}_{i,k}^{M} = \hat{u}_{i,k} + t_{n_t-n_r-1}^{\alpha}\sqrt{\hat{\sigma}_i^2 \boldsymbol{P}_{ii}}. \tag{6.74}$$

An occurrence of the fault for each i-th actuator is signaled when the input $u_{i,k}$ crosses the input adaptive threshold (cf. Fig. 6.17).

6.6 RUIF in Confidence Estimation of GMDH Model Inputs

The state-space GMDH neural model presented in Sect. 6.3 enables the robust fault detection with the application of the output adaptive thresholds but it does not allow to perform the fault isolation. The state-space representation of the GMDH network enables to develop a new RUIF-based approach in order to estimate the input signals of each neurons and the whole GMDH network. The calculation of the input adaptive thresholds for each input signal of the diagnosed system allows to perform the robust fault detection and the isolation of the actuators simultaneously. Let us consider a non-linear discrete-time system for the neuron model (6.50-6.51):

$$\boldsymbol{x}_{k+1} = \boldsymbol{A}\boldsymbol{x}_k + \boldsymbol{B}\boldsymbol{u}_k, \tag{6.75}$$

$$\boldsymbol{y}_{k+1} = \mathcal{G}(\boldsymbol{C}\boldsymbol{x}_{k+1}) + \boldsymbol{w}_{k+1}, \tag{6.76}$$

where $\boldsymbol{x}_k \in \mathbb{X} \subset \mathbb{R}^n$ is the state, $\boldsymbol{u}_k \in \mathbb{R}^{n_r}$ stands for the input, $\boldsymbol{y}_k \in \mathbb{R}^{n_s}$ denotes the output, and $\boldsymbol{w}_k \in l_2$ is the exogenous disturbance vector while:

$$l_2 = \{\boldsymbol{w} \in \mathbb{R}^n, \quad \|\boldsymbol{w}\|_{l_2} < +\infty\}, \tag{6.77}$$

where:

$$\|\boldsymbol{w}\|_{l_2} = \left(\sum_{k=0}^{\infty} \|\boldsymbol{w}_k\|^2\right)^{\frac{1}{2}}. \tag{6.78}$$

The system output can be written as follows:

$$\mathcal{G}^{-1}(\boldsymbol{y}_{k+1} - \boldsymbol{w}_{k+1}) = \boldsymbol{C}\boldsymbol{x}_{k+1} = \bar{\boldsymbol{y}}_{k+1}, \tag{6.79}$$

and

$$\mathcal{G}^{-1}(\boldsymbol{y}_{k+1} - \boldsymbol{w}_{k+1}) = \mathcal{G}^{-1}(\boldsymbol{y}_{k+1}) + \boldsymbol{v}_{k+1} = \bar{\boldsymbol{y}}_{k+1}, \tag{6.80}$$

$$\boldsymbol{H}[\mathcal{G}^{-1}(\boldsymbol{y}_{k+1}) + \boldsymbol{v}_{k+1}] = \boldsymbol{H}\boldsymbol{C}\boldsymbol{x}_{k+1}, \tag{6.81}$$

where $\boldsymbol{v}_k \in \mathcal{L}_2$. Substituting (6.75) into (6.81):

$$\boldsymbol{H}[\mathcal{G}^{-1}(\boldsymbol{y}_{k+1}) + \boldsymbol{v}_{k+1}] = \boldsymbol{H}\boldsymbol{C}\boldsymbol{A}\boldsymbol{x}_k + \boldsymbol{H}\boldsymbol{C}\boldsymbol{B}\boldsymbol{u}_k, \tag{6.82}$$

and assuming that:

$$\boldsymbol{H}\boldsymbol{C}\boldsymbol{B} = \boldsymbol{I}, \tag{6.83}$$

which implies that:

$$\operatorname{rank}(\boldsymbol{C}\boldsymbol{B}) = \operatorname{rank}(\boldsymbol{B}) = n_u, \tag{6.84}$$

the system input receives the following form:

$$\boldsymbol{u}_k = \boldsymbol{H}\mathcal{G}^{-1}(\boldsymbol{y}_{k+1}) + \boldsymbol{H}\boldsymbol{v}_{k+1} - \boldsymbol{H}\boldsymbol{C}\boldsymbol{A}\boldsymbol{x}_k. \tag{6.85}$$

On the basis of (6.85), the input estimate can be defined as:

$$\hat{\boldsymbol{u}}_k = \boldsymbol{H}\mathcal{G}^{-1}(\boldsymbol{y}_{k+1}) - \boldsymbol{H}\boldsymbol{C}\boldsymbol{A}\hat{\boldsymbol{x}}_k. \tag{6.86}$$

The state estimation error is given by:

$$\boldsymbol{e}_k = \boldsymbol{x}_k - \hat{\boldsymbol{x}}_k, \tag{6.87}$$

and the input estimation error can be defined as follows:

$$\varepsilon_{u,k} = \boldsymbol{u}_k - \hat{\boldsymbol{u}}_k = -\boldsymbol{H}\boldsymbol{C}\boldsymbol{A}\boldsymbol{e}_k + \boldsymbol{H}\boldsymbol{v}_{k+1}. \tag{6.88}$$

Taking into account the relation (6.79) the considered system (6.75-6.76) can be rewritten as:

$$\boldsymbol{x}_{k+1} = \boldsymbol{A}\boldsymbol{x}_k + \boldsymbol{B}\boldsymbol{u}_k, \tag{6.89}$$

$$\bar{\boldsymbol{y}}_{k+1} = \boldsymbol{C}\boldsymbol{x}_{k+1}. \tag{6.90}$$

Substituting (6.85) into (6.89):

$$\boldsymbol{x}_{k+1} = \boldsymbol{A}\boldsymbol{x}_k + \boldsymbol{B}\boldsymbol{H}\mathcal{G}^{-1}(\boldsymbol{y}_{k+1}) + \boldsymbol{B}\boldsymbol{H}\boldsymbol{v}_{k+1} - \boldsymbol{B}\boldsymbol{H}\boldsymbol{C}\boldsymbol{A}\boldsymbol{x}_k, \tag{6.91}$$

and assuming $\bar{\boldsymbol{A}} = \boldsymbol{A} - \boldsymbol{B}\boldsymbol{H}\boldsymbol{C}\boldsymbol{A}$ and $\bar{\boldsymbol{B}} = \boldsymbol{B}\boldsymbol{H}$, (6.89) receives the following form:

$$\boldsymbol{x}_{k+1} = \bar{\boldsymbol{A}}\boldsymbol{x}_k + \bar{\boldsymbol{B}}\mathcal{G}^{-1}(\boldsymbol{y}_{k+1}) + \bar{\boldsymbol{B}}\boldsymbol{v}_{k+1}. \tag{6.92}$$

The observer structure is:

$$\hat{\boldsymbol{x}}_{k+1} = \bar{\boldsymbol{A}}\hat{\boldsymbol{x}}_k + \bar{\boldsymbol{B}}\mathcal{G}^{-1}(\boldsymbol{y}_{k+1}) + \boldsymbol{K}(\mathcal{G}^{-1}(\boldsymbol{y}_k) - \boldsymbol{C}\hat{\boldsymbol{x}}_k), \tag{6.93}$$

while the state estimation error is given by:

$$\boldsymbol{e}_{k+1} = \boldsymbol{x}_{k+1} - \hat{\boldsymbol{x}}_{k+1} = \bar{\boldsymbol{A}}\boldsymbol{e}_k + \bar{\boldsymbol{B}}\boldsymbol{v}_{k+1} - \boldsymbol{K}(\mathcal{G}^{-1}(\boldsymbol{y}_k) - \boldsymbol{C}\hat{\boldsymbol{x}}_k). \tag{6.94}$$

If it is assumed that the system output can be defined as:

$$\boldsymbol{C}\boldsymbol{x}_k = \mathcal{G}^{-1}(\boldsymbol{y}_k - \boldsymbol{w}_k) = \mathcal{G}^{-1}(\boldsymbol{y}_k) + \boldsymbol{v}_k, \tag{6.95}$$

and from (6.95) it can be obtained:

$$\mathcal{G}^{-1}(\boldsymbol{y}_k) = \boldsymbol{C}\boldsymbol{x}_k - \boldsymbol{v}_k, \tag{6.96}$$

then substituting (6.96) into (6.94) the following form of the state estimation error is received:

$$\boldsymbol{e}_{k+1} = \bar{\boldsymbol{A}}\boldsymbol{e}_k + \bar{\boldsymbol{B}}\boldsymbol{v}_{k+1} - [\boldsymbol{K}(\boldsymbol{C}\boldsymbol{x}_k - \boldsymbol{v}_k - \boldsymbol{C}\hat{\boldsymbol{x}}_k)], \tag{6.97}$$

$$\boldsymbol{e}_{k+1} = \bar{\boldsymbol{A}}\boldsymbol{e}_k + \bar{\boldsymbol{B}}\boldsymbol{v}_{k+1} - \boldsymbol{K}\boldsymbol{C}\boldsymbol{e}_k + \boldsymbol{K}\boldsymbol{v}_k, \tag{6.98}$$

$$\boldsymbol{e}_{k+1} = (\bar{\boldsymbol{A}} - \boldsymbol{K}\boldsymbol{C})\boldsymbol{e}_k + \bar{\boldsymbol{B}}\boldsymbol{v}_{k+1} + \boldsymbol{K}\boldsymbol{v}_k, \tag{6.99}$$

and finally:

$$\boldsymbol{e}_{k+1} = \boldsymbol{A}_1\boldsymbol{e}_k + \bar{\boldsymbol{B}}\boldsymbol{v}_{k+1} + \boldsymbol{K}\boldsymbol{v}_k, \tag{6.100}$$

where: $\boldsymbol{A}_1 = \bar{\boldsymbol{A}} - \boldsymbol{K}\boldsymbol{C}$.

The objective is to design the observer in such a way that the state estimation error is asymptotically convergent and the following upper bound is guaranteed:

$$\|\varepsilon_{u,k}\|_{l_2} \leq \upsilon\|\boldsymbol{v}_k\|_{l_2}, \tag{6.101}$$

where $\upsilon > 0$ is the prescribed disturbance attenuation level. Thus, μ should be achieved with respect to the input estimation error but not to the state estimation error. Thus, the problem of $\mathcal{H}_\infty$ observer design [171] is to determine the gain matrix $\boldsymbol{K}$ such that:

$$\lim_{k\to\infty} \boldsymbol{e}_k = \boldsymbol{0} \quad \text{for} \quad \boldsymbol{v}_k = \boldsymbol{0}, \tag{6.102}$$

and

$$\|\varepsilon_{u,k}\|_{l_2} \leq \upsilon\|\boldsymbol{v}\|_{l_2} \quad \text{for} \quad \boldsymbol{v}_k \neq \boldsymbol{0} \quad \text{and} \quad \boldsymbol{e}_0 = \boldsymbol{0}. \tag{6.103}$$

In order to settle the above problem it is sufficient to find the Lyapunov function V_k such that:

$$\Delta\boldsymbol{V} + \varepsilon_{u,k}^T\varepsilon_{u,k} - \mu^2\boldsymbol{v}_{k+1}^T\boldsymbol{v}_{k+1} - \mu^2\boldsymbol{v}_k^T\boldsymbol{v}_k < 0, \tag{6.104}$$

where $\Delta V_k = V_{k+1} - V_k$, $v_k = \boldsymbol{e}_k^T\boldsymbol{P}\boldsymbol{e}_k$ and $\mu > 0$.

Indeed, if $\boldsymbol{v}_k = \boldsymbol{0}$ for $k = 0, \ldots, \infty$ then (6.104) boils down to:

$$\Delta V_k + \varepsilon_{u,k}^T\varepsilon_{u,k} < 0 \quad \text{for} \quad k = 0, \ldots\infty, \tag{6.105}$$

and hence $\Delta V_k < 0$, which leads to (6.102). If $\boldsymbol{v}_k \neq \boldsymbol{0}$, $(k = 0, \ldots, \infty)$ then inequality (6.104) yields:

$$J = \sum_{k=0}^{\infty}\left(\Delta V_k + \varepsilon_{u,k}^T\varepsilon_{u,k} - \mu^2\boldsymbol{v}_k^T\boldsymbol{v}_k - \mu^2\boldsymbol{v}_{k+1}\boldsymbol{v}_{k+1}\right) < 0, \tag{6.106}$$

which can be written as:

$$J = V_\infty - V_0 + \sum_{k=0}^{\infty}\varepsilon_{u,k}^T\varepsilon_{u,k} - \mu^2\sum_{k=0}^{\infty}\boldsymbol{v}_k^T\boldsymbol{v}_k - \mu^2\sum_{k=0}^{\infty}\boldsymbol{v}_{k+1}^T\boldsymbol{v}_{k+1} < 0. \tag{6.107}$$

Bearing in mind that:

$$\mu^2 \sum_{k=0}^{\infty} v_{k+1}^T v_{k+1} = \mu^2 \sum_{k=0}^{\infty} v_k^T v_k - \mu^2 v_0^T v_0, \tag{6.108}$$

inequality (6.107) can be written as:

$$J = V_\infty - V_0 + \sum_{k=0}^{\infty} \varepsilon_{u,k}^T \varepsilon_{u,k} - 2\mu^2 \sum_{k=0}^{\infty} v_k^T v_k + \mu^2 v_0^T v_0 < 0. \tag{6.109}$$

Knowing that $V_0 = 0$ for $\boldsymbol{e}_0 = 0$ and $V_\infty \geq 0$, (6.109) leads to (6.103) with $v = \sqrt{2}\mu$.

Since the general framework for designing the robust observer is given, then the following form of the Lyapunov function is proposed [181]:

$$\Delta V = \boldsymbol{e}_{k+1}^T \boldsymbol{P} \boldsymbol{e}_{k+1} - \boldsymbol{e}_k^T \boldsymbol{P} \boldsymbol{e}_k, \tag{6.110}$$

and hence:

$$\begin{aligned}
\boldsymbol{e}_{k+1}^T \boldsymbol{P} \boldsymbol{e}_{k+1} =& [\boldsymbol{e}_k^T \boldsymbol{A}_1^T + v_{k+1}^T \bar{\boldsymbol{B}}^T + v_k^T \boldsymbol{K}^T] \boldsymbol{P} [\boldsymbol{A}_1 \boldsymbol{e}_k + \bar{\boldsymbol{B}} v_{k+1} + \boldsymbol{K} v_k] = \\
& \boldsymbol{e}_k^T \boldsymbol{A}_1^T \boldsymbol{P} \boldsymbol{A}_1 \boldsymbol{e}_k + \boldsymbol{e}_k^T \boldsymbol{A}_1^T \boldsymbol{P} \bar{\boldsymbol{B}} v_{k+1} + \boldsymbol{e}_k^T \boldsymbol{A}_1^T \boldsymbol{P} \boldsymbol{K} v_k + \\
& v_{k+1}^T \bar{\boldsymbol{B}}^T \boldsymbol{P} \boldsymbol{A}_1 \boldsymbol{e}_k + v_{k+1}^T \bar{\boldsymbol{B}}^T \boldsymbol{P} \bar{\boldsymbol{B}} v_{k+1} + v_{k+1}^T \bar{\boldsymbol{B}}^T \boldsymbol{P} \boldsymbol{K} v_k + \\
& v_k^T \boldsymbol{K}^T \boldsymbol{P} \boldsymbol{A}_1 \boldsymbol{e}_k + v_k^T \boldsymbol{K}^T \boldsymbol{P} \bar{\boldsymbol{B}} v_{k+1} + v_k^T \boldsymbol{K}^T \boldsymbol{P} \boldsymbol{K} v_k,
\end{aligned} \tag{6.111}$$

$$\begin{aligned}
\varepsilon_{u,k}^T \varepsilon_{u,k} =& [-\boldsymbol{e}_k^T \boldsymbol{A}^T \boldsymbol{C}^T \boldsymbol{H}^T + v_{k+1}^T \boldsymbol{H}^T][-\boldsymbol{H}\boldsymbol{C}\boldsymbol{A}\boldsymbol{e}_k + \boldsymbol{H} v_{k+1}] = \\
& [-\boldsymbol{e}_k^T \boldsymbol{A}^T \boldsymbol{C}^T + v_{k+1}^T] \boldsymbol{H}^T \boldsymbol{H} [-\boldsymbol{C}\boldsymbol{A}\boldsymbol{e}_k + v_{k+1}] = \\
& \boldsymbol{e}_k^T \boldsymbol{A}^T \boldsymbol{C}^T \boldsymbol{H}^T \boldsymbol{H}\boldsymbol{C}\boldsymbol{A}\boldsymbol{e}_k - \boldsymbol{e}_k^T \boldsymbol{A}^T \boldsymbol{C}^T \boldsymbol{H}^T \boldsymbol{H} v_{k+1} - \\
& v_{k+1}^T \boldsymbol{H}^T \boldsymbol{H}\boldsymbol{C}\boldsymbol{A}\boldsymbol{e}_k + v_{k+1}^T \boldsymbol{H}^T \boldsymbol{H} v_{k+1}.
\end{aligned} \tag{6.112}$$

Thus, for $\boldsymbol{z}_k = [\boldsymbol{e}_k, v_k, v_{k+1}]^T$ the inequality (6.104) becomes:

$$\boldsymbol{z}_k^T \boldsymbol{X} \boldsymbol{z}_k < 0, \tag{6.113}$$

where the matrix $\boldsymbol{X} \prec 0$ has the following form:

$$\begin{bmatrix}
\boldsymbol{A}_1^T \boldsymbol{P} \boldsymbol{A}_1 - \boldsymbol{P} + \boldsymbol{A}^T \boldsymbol{C}^T \boldsymbol{H}^T \boldsymbol{H}\boldsymbol{C}\boldsymbol{A} & \boldsymbol{A}_1^T \boldsymbol{P} \boldsymbol{K} & \boldsymbol{A}_1^T \boldsymbol{P} \bar{\boldsymbol{B}} - \boldsymbol{A}^T \boldsymbol{C}^T \boldsymbol{H}^T \boldsymbol{H} \\
\boldsymbol{K}^T \boldsymbol{P} \boldsymbol{A}_1 & \boldsymbol{K}^T \boldsymbol{P} \boldsymbol{K} - \mu^2 \boldsymbol{I} & \boldsymbol{K}^T \boldsymbol{P} \bar{\boldsymbol{B}} \\
\bar{\boldsymbol{B}}^T \boldsymbol{P} \boldsymbol{A}_1 - \boldsymbol{H}^T \boldsymbol{H}\boldsymbol{C}\boldsymbol{A} & \bar{\boldsymbol{B}}^T \boldsymbol{P} \boldsymbol{K} & \bar{\boldsymbol{B}}^T \boldsymbol{P} \bar{\boldsymbol{B}} + \boldsymbol{H}^T \boldsymbol{H} - \mu^2 \boldsymbol{I}
\end{bmatrix}. \tag{6.114}$$

Observing that:

$$\begin{bmatrix} \boldsymbol{A}_1^T \\ \boldsymbol{K}^T \\ \bar{\boldsymbol{B}}^T \end{bmatrix} \boldsymbol{P} \begin{bmatrix} \boldsymbol{A}_1 \\ \boldsymbol{K} \\ \bar{\boldsymbol{B}} \end{bmatrix}^T = \begin{bmatrix}
\boldsymbol{A}_1^T \boldsymbol{P} \boldsymbol{A}_1 & \boldsymbol{A}_1^T \boldsymbol{P} \boldsymbol{K} & \boldsymbol{A}_1^T \boldsymbol{P} \bar{\boldsymbol{B}} \\
\boldsymbol{K}^T \boldsymbol{P} \boldsymbol{A}_1 & \boldsymbol{K}^T \boldsymbol{P} \boldsymbol{K} & \boldsymbol{K}^T \boldsymbol{P} \bar{\boldsymbol{B}} \\
\bar{\boldsymbol{B}}^T \boldsymbol{P} \boldsymbol{A}_1 & \bar{\boldsymbol{B}}^T \boldsymbol{P} \boldsymbol{K} & \bar{\boldsymbol{B}}^T \boldsymbol{P} \bar{\boldsymbol{B}}
\end{bmatrix}, \tag{6.115}$$

equation (6.114) can be rewritten as:

$$\begin{bmatrix} \boldsymbol{A}_1^T \\ \boldsymbol{K}^T \\ \bar{\boldsymbol{B}} \end{bmatrix} \boldsymbol{P} \begin{bmatrix} \boldsymbol{A}_1^T \\ \boldsymbol{K}^T \\ \bar{\boldsymbol{B}} \end{bmatrix}^T + \begin{bmatrix} -\boldsymbol{P} + \boldsymbol{A}^T\boldsymbol{C}^T\boldsymbol{H}^T\boldsymbol{H}\boldsymbol{C}\boldsymbol{A} & \boldsymbol{0} & -\boldsymbol{A}^T\boldsymbol{C}^T\boldsymbol{H}^T\boldsymbol{H} \\ \boldsymbol{0} & -\mu^2\boldsymbol{I} & \boldsymbol{0} \\ -\boldsymbol{H}^T\boldsymbol{H}\boldsymbol{C}\boldsymbol{A} & \boldsymbol{0} & \boldsymbol{H}^T\boldsymbol{H} - \mu^2\boldsymbol{I} \end{bmatrix} < \boldsymbol{0}. \quad (6.116)$$

Moreover, applying the Schur complements, (6.116) is equivalent to (6.117):

$$\begin{bmatrix} -\boldsymbol{P} + \boldsymbol{A}^T\boldsymbol{C}^T\boldsymbol{H}^T\boldsymbol{H}\boldsymbol{C}\boldsymbol{A} & \boldsymbol{0} & -\boldsymbol{A}^T\boldsymbol{C}^T\boldsymbol{H}^T\boldsymbol{H} & \boldsymbol{A}_1^T \\ \boldsymbol{0} & -\mu^2\boldsymbol{I} & \boldsymbol{0} & \boldsymbol{K}^T \\ -\boldsymbol{H}^T\boldsymbol{H}\boldsymbol{C}\boldsymbol{A} & \boldsymbol{0} & \boldsymbol{H}^T\boldsymbol{H} - \mu^2\boldsymbol{I} & \bar{\boldsymbol{B}} \\ \boldsymbol{A}_1 & \boldsymbol{K} & \bar{\boldsymbol{B}} & -\boldsymbol{P}^{-1} \end{bmatrix} < \boldsymbol{0}. \quad (6.117)$$

Multiplying (6.117) from both sites by:

$$\mathrm{diag}(\boldsymbol{I}, \boldsymbol{I}, \boldsymbol{I}, \boldsymbol{P}), \quad (6.118)$$

(6.117) receives the form:

$$\begin{bmatrix} -\boldsymbol{P} + \boldsymbol{A}^T\boldsymbol{C}^T\boldsymbol{H}^T\boldsymbol{H}\boldsymbol{C}\boldsymbol{A} & \boldsymbol{0} & -\boldsymbol{A}^T\boldsymbol{C}^T\boldsymbol{H}^T\boldsymbol{H} & \boldsymbol{A}_1^T\boldsymbol{P} \\ \boldsymbol{0} & -\mu^2\boldsymbol{I} & \boldsymbol{0} & \boldsymbol{K}^T\boldsymbol{P} \\ -\boldsymbol{H}^T\boldsymbol{H}\boldsymbol{C}\boldsymbol{A} & \boldsymbol{0} & \boldsymbol{H}^T\boldsymbol{H} - \mu^2\boldsymbol{I} & \bar{\boldsymbol{B}}\boldsymbol{P} \\ \boldsymbol{P}\boldsymbol{A}_1 & \boldsymbol{P}\boldsymbol{K} & \boldsymbol{P}\bar{\boldsymbol{B}} & -\boldsymbol{P} \end{bmatrix} < \boldsymbol{0}, \quad (6.119)$$

and then substituting:

$$\boldsymbol{A}_1 = \bar{\boldsymbol{A}} - \boldsymbol{K}\boldsymbol{C}, \quad (6.120)$$

$$\boldsymbol{P}\boldsymbol{A}_1 = \boldsymbol{P}\bar{\boldsymbol{A}} - \boldsymbol{P}\boldsymbol{K}\boldsymbol{C} = \boldsymbol{P}\bar{\boldsymbol{A}} - \boldsymbol{N}\boldsymbol{C}, \quad (6.121)$$

$$\boldsymbol{N} = \boldsymbol{P}\boldsymbol{K}, \quad (6.122)$$

where: $\boldsymbol{N} \in \mathbb{R}^{n \times m}$

$$\boldsymbol{A}_1^T\boldsymbol{P} = \bar{\boldsymbol{A}}^T\boldsymbol{P} - \boldsymbol{C}^T\boldsymbol{N}^T, \quad (6.123)$$

equation (6.117) receives the form:

$$\begin{bmatrix} -\boldsymbol{P} + \boldsymbol{A}^T\boldsymbol{C}^T\boldsymbol{H}^T\boldsymbol{H}\boldsymbol{C}\boldsymbol{A} & \boldsymbol{0} & -\boldsymbol{A}^T\boldsymbol{C}^T\boldsymbol{H}^T\boldsymbol{H} & \bar{\boldsymbol{A}}^T\boldsymbol{P} - \boldsymbol{C}^T\boldsymbol{N}^T \\ \boldsymbol{0} & -\mu^2\boldsymbol{I} & \boldsymbol{0} & \boldsymbol{N}^T \\ -\boldsymbol{H}^T\boldsymbol{H}\boldsymbol{C}\boldsymbol{A} & \boldsymbol{0} & \boldsymbol{H}^T\boldsymbol{H} - \mu^2\boldsymbol{I} & \bar{\boldsymbol{B}}^T\boldsymbol{P} \\ \boldsymbol{P}\bar{\boldsymbol{A}} - \boldsymbol{N}\boldsymbol{C} & \boldsymbol{N} & \boldsymbol{P}\bar{\boldsymbol{B}} & -\boldsymbol{P} \end{bmatrix} < \boldsymbol{0}. \quad (6.124)$$

Note that (6.124) is a usual Linear Matrix Inequality (LMI), which can be solved, e.g., with the Matlab. As the result for the given disturbance attenuation level μ the observer gain matrix $\boldsymbol{K}$ and the estimate of the inputs $\hat{\boldsymbol{u}}_k$ can be obtained.

The presented above approach allows to obtain the estimates of the GMDH neural networks inputs. Moreover, on the basis of (6.104):

$$\varepsilon_{u,k}^T\varepsilon_{u,k} \leq \mu^2 v_{k+1}^T v_{k+1} + \mu^2 v_k^T v_k. \quad (6.125)$$

Assuming that $v_k^T v_k = \|v_k\|_2^2 < \delta$, where $\delta > 0$ is a given bound then:

$$\varepsilon_{u,k}^T \varepsilon_{u,k} \leq 2\mu^2\delta, \tag{6.126}$$

the adaptive thresholds for the inputs of the GMDH neural model receive the following form:

$$\hat{u}_{i,k} - \mu\sqrt{2\delta} \leq u_{i,k} \leq \hat{u}_{i,k} + \mu\sqrt{2\delta}. \tag{6.127}$$

The occurrence of the fault for each i-th actuator is signaled when input $u_{i,k}$ crosses the input adaptive threshold presented in Fig. (6.18).

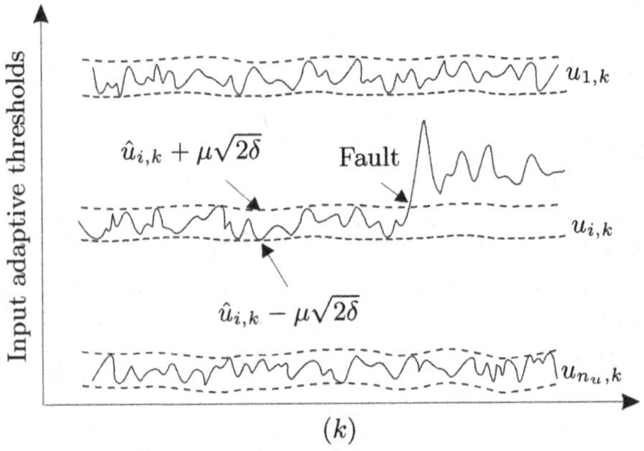

Fig. 6.18. Input adaptive thresholds obtained via the RUIF

6.7 An Illustrative Example – UKF and RUIF in Confidence Estimation

Let us consider a non-linear system:

$$\boldsymbol{x}_{k+1} = \boldsymbol{A}\boldsymbol{x}_k + \boldsymbol{B}\boldsymbol{u}_k, \tag{6.128}$$
$$\boldsymbol{y}_{k+1} = \mathcal{G}(\boldsymbol{C}\boldsymbol{x}_{k+1}) + \boldsymbol{w}_{k+1}, \tag{6.129}$$

with:

$$\boldsymbol{A} = \begin{bmatrix} 0.1 & 0.2 \\ 0 & 0.3 \end{bmatrix}, \quad \boldsymbol{B} = \begin{bmatrix} 1 & 2 \\ 3 & 4 \end{bmatrix}, \quad \boldsymbol{C} = \begin{bmatrix} 1 & 0 \\ 0 & 1 \end{bmatrix},$$

while the input are generated with uniform distribution: $\boldsymbol{u}_k = \mathcal{U}(0.1, 0.1)$, for $k = 1, ..., 100$, and the exogenous disturbance is $\boldsymbol{w}_k \sim \mathcal{N}(\boldsymbol{0}, 0.01\boldsymbol{I}_n)$. Moreover, the non-linear function is assumed as $\mathcal{G}(\cdot) = \tanh(\cdot)$. On the basis of the data set $\{\boldsymbol{u}_k, \boldsymbol{y}_k\}$ for $k = 1, ..., 100$ the estimates of the matrixes $\hat{\boldsymbol{A}}$ and $\hat{\boldsymbol{B}}$ are obtained with the application of the UKF-based approach described in Sect. 6.3.

$$\hat{\boldsymbol{A}} = \begin{bmatrix} 0.0971 & 0.2021 \\ 0 & 0.3054 \end{bmatrix}, \quad \hat{\boldsymbol{B}} = \begin{bmatrix} 1.0063 & 2.0234 \\ 3.0179 & 4.0490 \end{bmatrix}.$$

Figures 6.19 and 6.20 show the system responses generated according to eqs. (6.128)-(6.129) and the responses for the $\hat{\boldsymbol{A}}$ and $\hat{\boldsymbol{B}}$ obtained with the UKF. The sum squared error for the system responses $y_{1,k}$ and $y_{2,k}$ are $Q_{SSE} = 0.0010$ and $Q_{SSE} = 0.0011$, respectively.

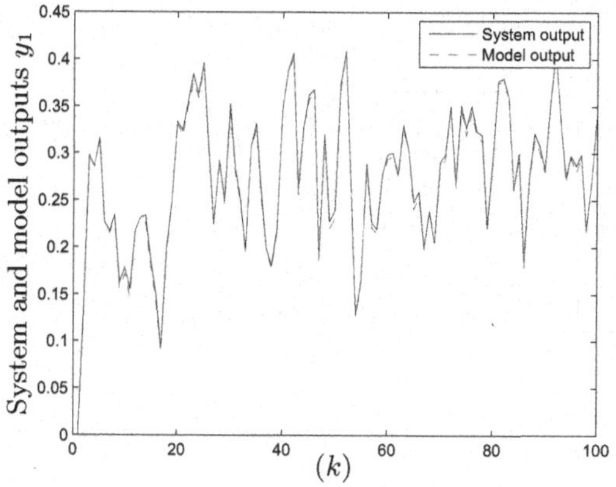

Fig. 6.19. System and model outputs $y_{1,k}$ obtained via the UKF

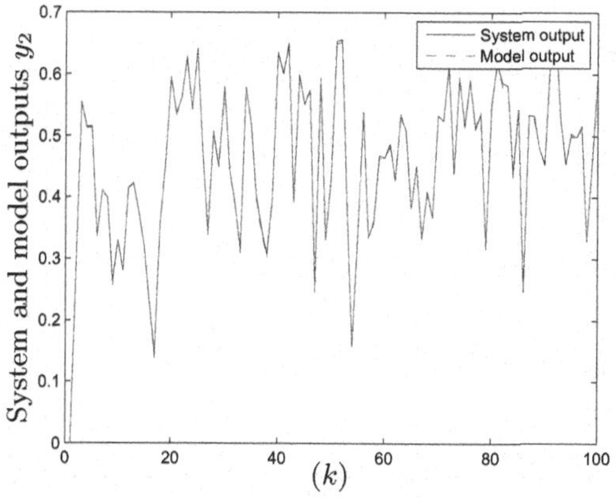

Fig. 6.20. System and model outputs $y_{2,k}$ obtained via the UKF

At the next stage the methodology presented in Sect. 6.6 is applied in order to estimate the inputs $u_{1,k}$ and $u_{2,k}$ for the matrixes $\hat{\boldsymbol{A}}$ and $\hat{\boldsymbol{B}}$ obtained with the UKF. According to the presented methodology the following gain matrix $\boldsymbol{K}$ for $\mu = 2.769$ is obtained:

$$\boldsymbol{K} = \begin{bmatrix} -0.7990 & -0.2543 \\ 0.3552 & 0.1114 \end{bmatrix}. \tag{6.130}$$

Let the initial condition for the system and the observer be: $\boldsymbol{x}_0 = [0.5, 0.5]^T$ and $\hat{\boldsymbol{x}}_0 = \boldsymbol{0}$ while the exogenous disturbance are $\boldsymbol{w}_k \sim \mathcal{N}(\boldsymbol{0}, 0.01\boldsymbol{I}_n)$ and the inputs are generated as follows:

$$\boldsymbol{u}_{1,k} = 0.2\sin(0.2\pi k), \tag{6.131}$$

$$\boldsymbol{u}_{2,k} = 0.1\sin(0.2\pi k/5). \tag{6.132}$$

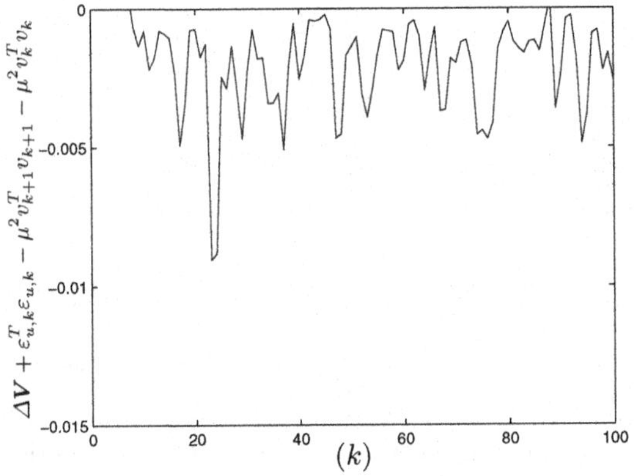

Fig. 6.21. Evolution of $\Delta V + \varepsilon_{u,k}^T \varepsilon_{u,k} - \mu^2 v_{k+1}^T v_{k+1} - \mu^2 v_k^T v_k$

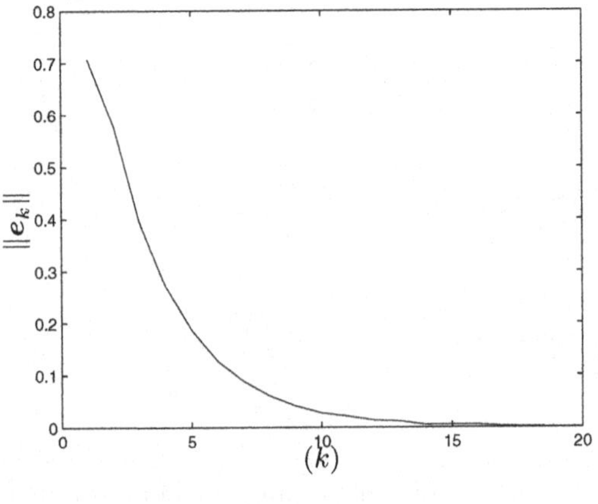

Fig. 6.22. Evolution of $\|\boldsymbol{e}_k\|$

First, let us consider the case when $\hat{\boldsymbol{x}}_0 = \boldsymbol{x}_0$, $(\boldsymbol{e}_0 = \mathbf{0})$. Figure 6.21 clearly indicates that the condition (6.103) is satisfied, which means that the attenuation level $\mu = 2.769$ is achieved.

Now let us assume that $v_k = \mathbf{0}$ and $\hat{\boldsymbol{x}}_0 \neq \boldsymbol{x}_0$. Figure 6.22 clearly shows that (6.102) is satisfied as well. Moreover, Fig. 6.23 depicts the convergence of the norm of the input estimation error equal to zero.

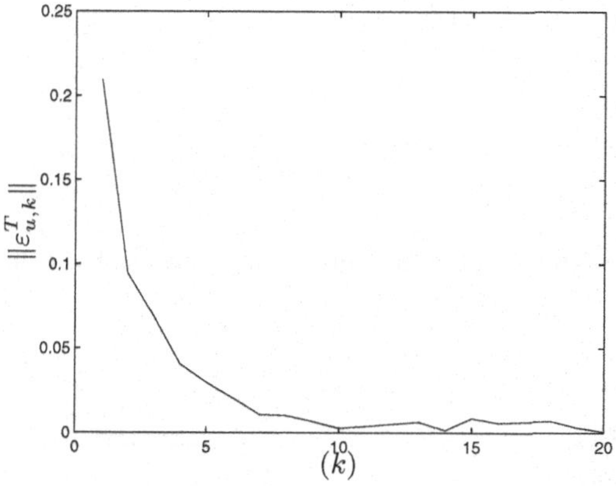

Fig. 6.23. Evolution of $\|\varepsilon_{u,k}^T\|$

Figures 6.24 and 6.25 present the system and model responses $y_{1,k}$ and $y_{2,k}$ obtained for the input signals (6.131) with the application of the RUIF.

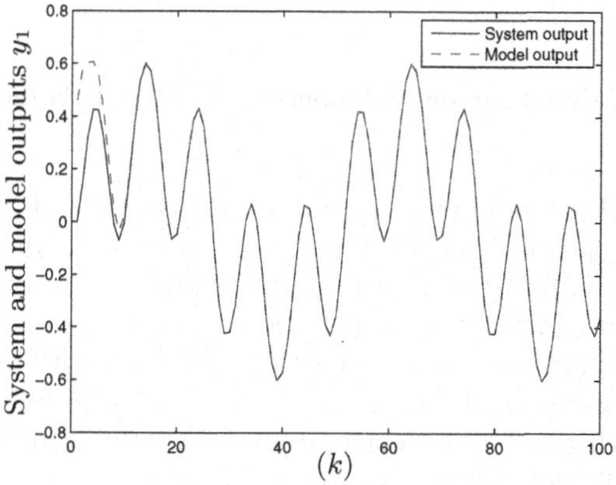

Fig. 6.24. System and model outputs $y_{1,k}$ obtained via the RUIF

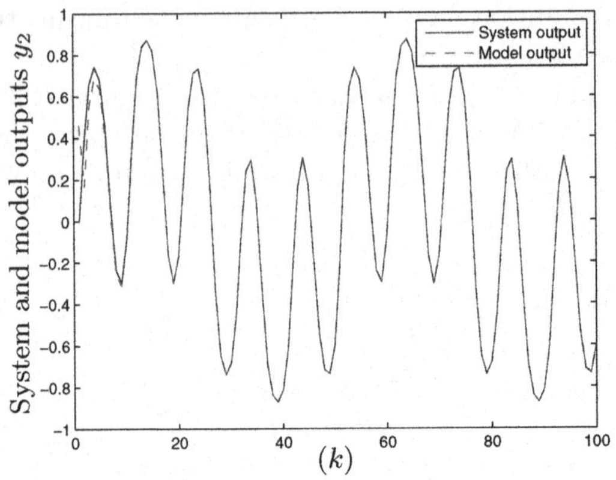

Fig. 6.25. System and model outputs $y_{2,k}$ obtained via the RUIF

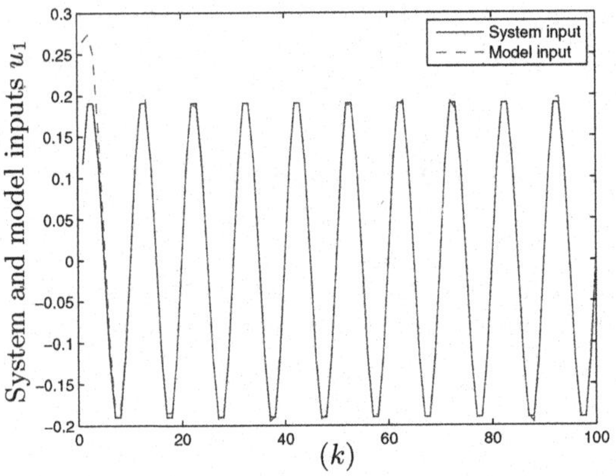

Fig. 6.26. System and model inputs $u_{1,k}$ obtained via the RUIF

The system and model inputs $u_{1,k}$ and $u_{2,k}$ obtained with the application of the RUIF approach is shown at Figs. 6.26 and 6.27. As it can be noticed in Figs. 6.22 and 6.23 the proposed approach provides an efficient way for increasing the decay rate of the state and input estimation errors. Moreover, Figs. 6.24-6.27 confirm the high quality of modelling with the application of the proposed approach. From these results, it can be observed that the proposed approach can be efficiently applied for solving robust $\mathcal{H}_\infty$-based identification of the non-linear systems.

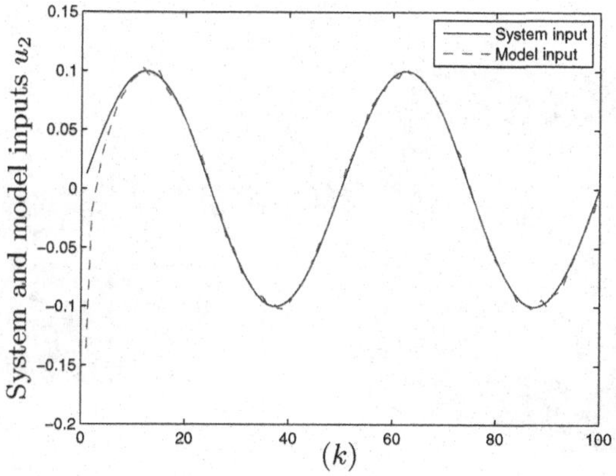

Fig. 6.27. System and model inputs $u_{2,k}$ obtained via the RUIF

6.8 State-Space GMDH Model in Robust FDI of Tunnel Furnace

The objective of this section is to design a dynamic GMDH model according to the approach described in Sect. 6.3 and apply it to the robust fault detection of a tunnel furnace (cf. Figs. 6.28-6.29).

The considered tunnel furnace is designed to mimic, in the laboratory conditions, the real industrial tunnel furnaces, which can be applied in the food industry or production of ceramics among others. The furnace is equipped in three electric heaters and four temperature sensors. The required temperature of the furnace can be kept by controlling the heaters behaviour. This task can be achieved by a group regulation of voltage with the application of the controller PACSystems RX3i manufactured by GE Fanuc Intelligent Platforms and semiconductor relays RP6 produced by LUMEL providing an impulse control with a variable impulse frequency $fmax = 1\ Hz$. The maximum power outputs of the heaters were measured to be approximately $686W$, $693W$ and $756W \pm 20W$, respectively.

The temperature of the furnace is measured via IC695ALG600 module with Pt100 Resistive Thermal Devices (RTDs) with an accuracy of $\pm 0.7°C$. The visualisation of the behaviour of the tunnel furnace is made by Quick-panel CE device from GE Fanuc Intelligent Platforms. It is worth to note that the considered system is a distributedparameter one (i.e., a system whose state space is infinite dimensional), thus any resulting model from input-output data will be at best an approximation.

The tunnel furnace can be considered as a three-input and four-output system $(t_1, t_2, t_3, t_4) = f(u_1, u_2, u_3)$, where $t_1, \ldots, t_4$ represent measurements of temperatures from four subsequent sensors and values $u_1, \ldots, u_3$ denote

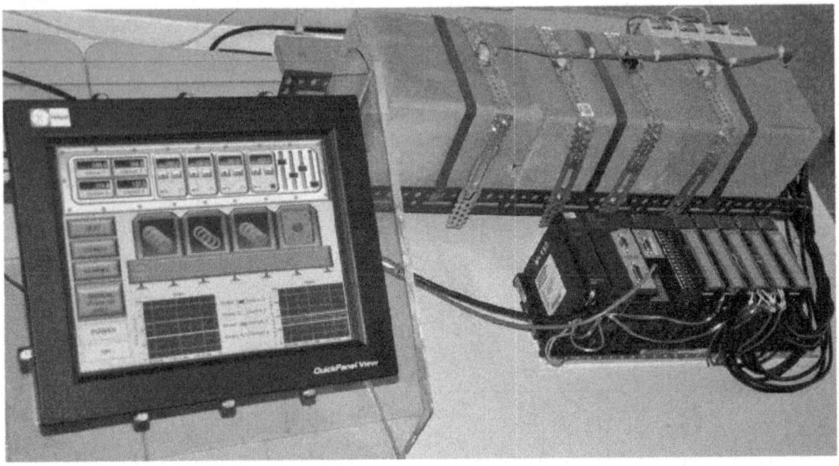

Fig. 6.28. Laboratory model of a tunnel furnace

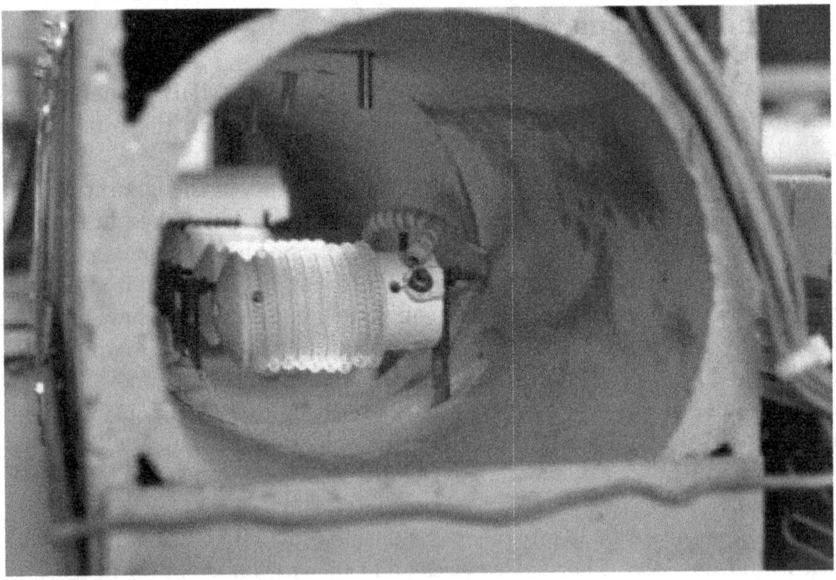

Fig. 6.29. Interior of a tunnel furnace

the input voltages allowing to control the heaters. The input and output data used for the identification and validation are collected in two data sets consisting of 2600-th samples. It should be also pointed out that these data sets are scaled for the purpose of neural networks designing. The output data signals should be transformed taking into consideration the response range of the output neurons. For the hyperbolic tangent neurons this range is $[-1, 1]$. To perform such a kind of transformation, the linear scaling can be used. Moreover, the data sets used for the identification and validation are filtered with application of the second order Butterworth filters. Furthermore, the offset levels from data sets are removed.

The parameters of the state-space dynamic neurons are estimated with the application of the training algorithm, which is based on the UKF approach presented in Sect. 6.4. The selection of the best performing neurons in the terms of their processing accuracy is realized with the application of the SSM based on the sum squared error evaluation criterion. Table 6.1 presents the values of the evolution criterion $Q_{\mathcal{V}}(\hat{t}_1) - Q_{\mathcal{V}}(\hat{t}_4)$ for the best performing neurons in a particular layer of the GMDH neural network.

Table 6.1. Values of evaluation criterion for the best neurons in the subsequent layers of the network for the validation data

Layer	$Q_{\mathcal{V}}(\hat{t}_1)$	$Q_{\mathcal{V}}(\hat{t}_2)$	$Q_{\mathcal{V}}(\hat{t}_3)$	$Q_{\mathcal{V}}(\hat{t}_4)$
1	0.0174	0.0154	0.0401	0.0212
2	0.0321	0.0136	0.0329	0.0188
3	0.0007	0.0006	0.0091	0.0012
4	0.0005	0.0003	0.0007	0.0009
5	0.0071	0.0010	0.0141	0.0054

Additionally, for the sake of a comparison, the results obtained with the application of the linear state-space model are presented (cf. Tab. 6.2). In particular, as a result of using a models within the range from 1 up to 10, the 5-th order state-space model is applied.

Table 6.2. Values of evaluation criterion for the non-linear dynamic GMDH and linear state-space models for the validation data

Model	$Q_{\mathcal{V}}(\hat{t}_1)$	$Q_{\mathcal{V}}(\hat{t}_2)$	$Q_{\mathcal{V}}(\hat{t}_3)$	$Q_{\mathcal{V}}(\hat{t}_4)$
Linear	0.0060	0.0037	0.0040	0.0027
GMDH	0.0005	0.0003	0.0007	0.0009

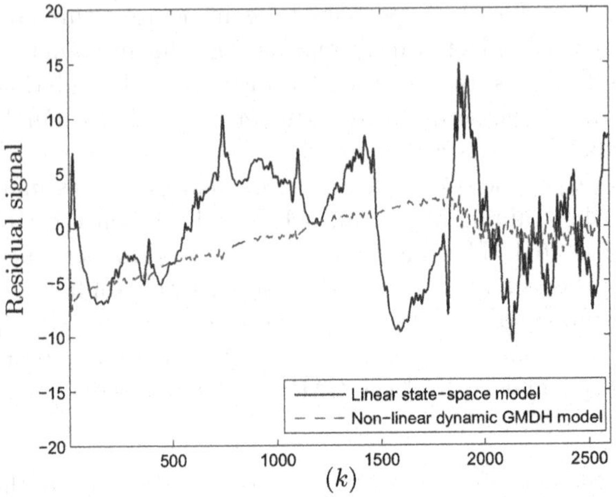

Fig. 6.30. Residuals of temperature t_1 for the linear state-space and non-linear GMDH models

Figure 6.30 presents the residuals signals calculated as the difference between temperature t_1 of the tunnel furnace and the linear state-space model and the non-linear dynamic GMDH model, respectively. The results show that the quality of the non-linear dynamic GMDH model is better than the linear state-space model. It follows from non-linear nature of the identified tunnel furnace.

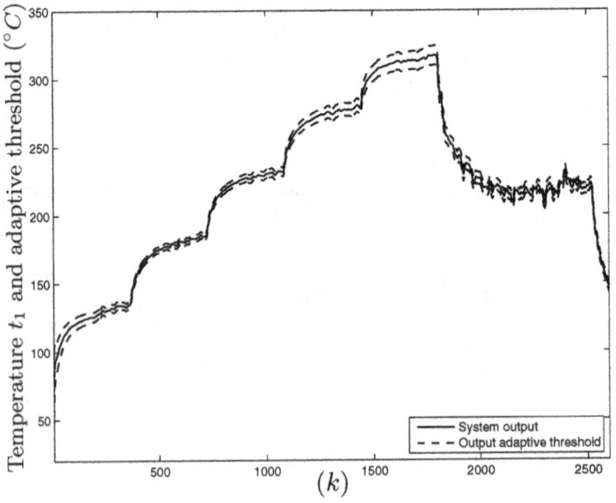

Fig. 6.31. Temperature t_1 and output adaptive threshold

Figures 6.31–6.34 depict temperatures $t_1 - t_4$ of the furnace and the adaptive thresholds obtained with (6.48-6.49) for the validation data set (no fault case).

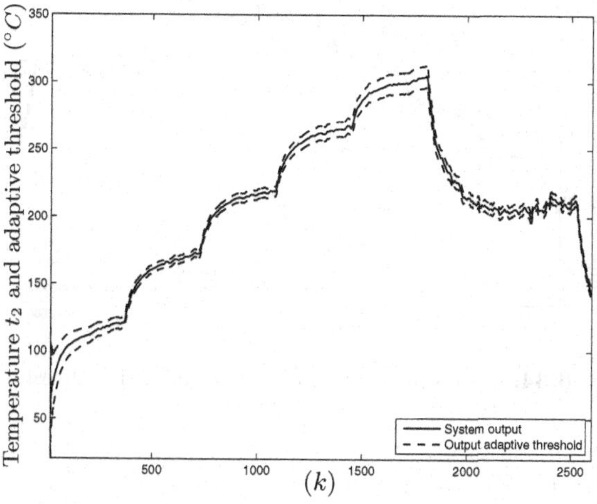

Fig. 6.32. Temperature t_2 and output adaptive threshold

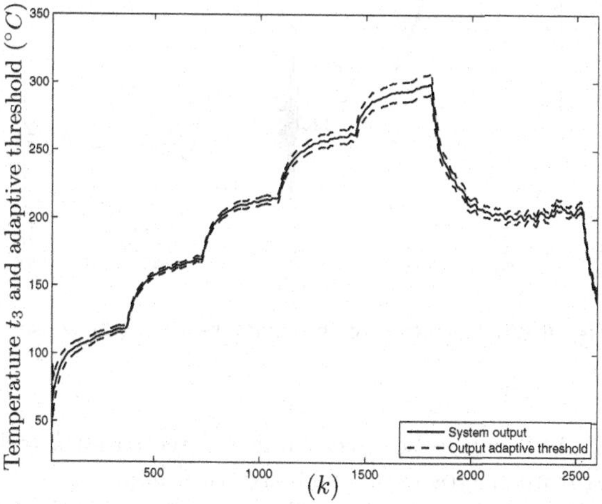

Fig. 6.33. Temperature t_3 and output adaptive threshold

After the synthesis of the GMDH model, it is possible to employ it for the robust fault detection of the tunnel furnace. Figure 6.35 presents the measurements of temperature t_1 from the faulty sensor (simulated during 10 secs.) and the adaptive threshold obtained with the application of the GMDH

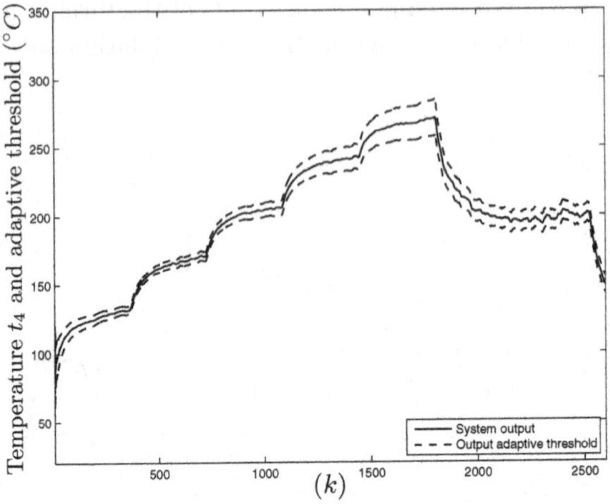

Fig. 6.34. Temperature t_4 and output adaptive threshold

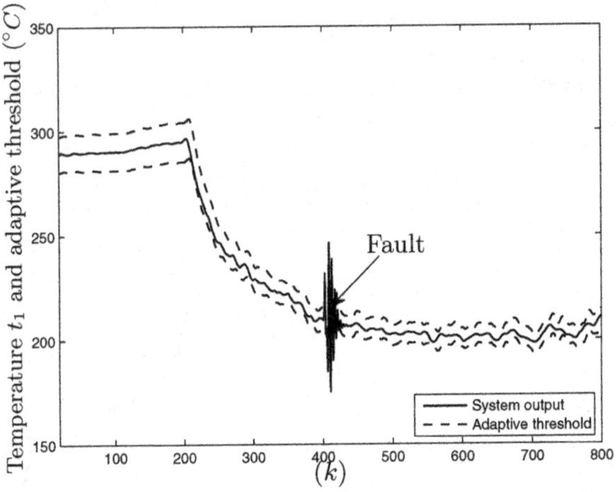

Fig. 6.35. Detection of the faulty temperature sensor

neural network. As it can be seen the fault is detected for $k = 400$ when the value of temperature t_1 crosses the adaptive threshold.

At the next stage of the experiment the input signals of the GMDH neural model are estimated with the application of the approach based on the application of the RUIF presented in Sect. 6.6. The input voltages u_1–u_3 of the electric heaters and the corresponding input adaptive thresholds are given in Figs. 6.36–6.38, respectively.

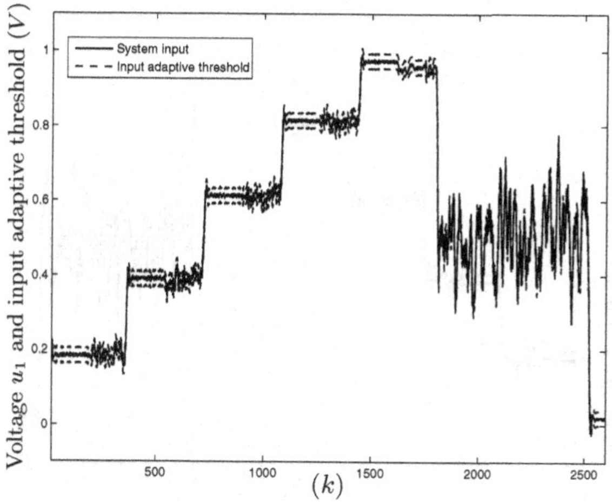

Fig. 6.36. Voltage u_1 and input adaptive threshold

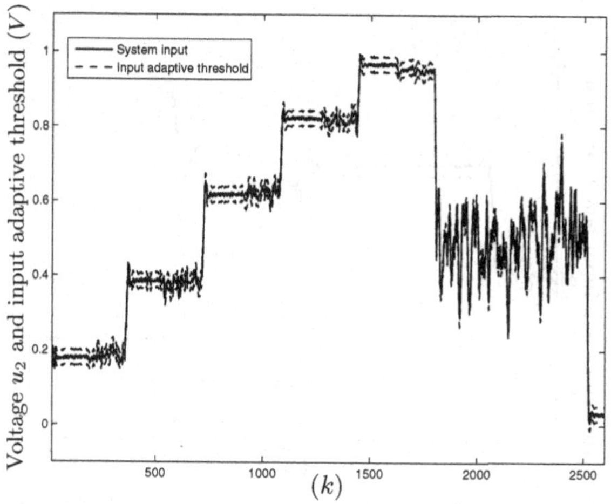

Fig. 6.37. Voltage u_2 and input adaptive threshold

Figure 6.39 presents the measurements of the input voltage u_1 and the corresponding input adaptive threshold obtained with the application of the GMDH neural network and RUIF. As it can be seen the fault is detected for $k = 300$ when the value of voltage u_1 crosses the input adaptive threshold.

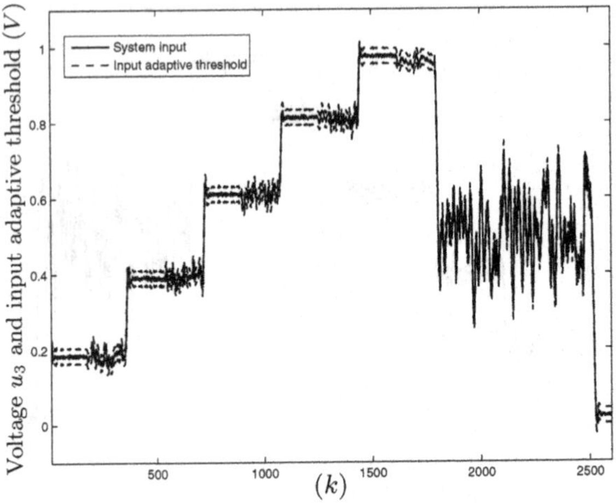

Fig. 6.38. Voltage u_3 and input adaptive threshold

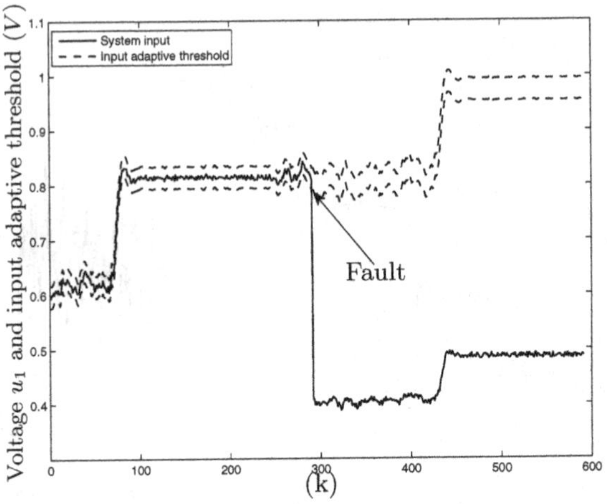

Fig. 6.39. Detection of the faulty electric heater

6.9 Concluding Remarks

The objective of this chapter is concerned with the design of the robust fault detection and actuators fault localisation system based on the dynamic neural model. To tackle this problem, the GMDH neural network consisting of the dynamic MIMO neural neurons in the state-space representation was developed. The state-space representation allows to define the stability conditions for each dynamic neuron and the whole GMDH network. Taking into account these conditions during the parameters estimation with the application of the

UKF the stable dynamic non-linear GMDH neural model can be obtained. Moreover, the application of the UKF allows to obtain the uncertainty of the GMDH neural model. This knowledge enables to calculate the output adaptive threshold of the GMDH model and apply it to the robust fault detection of the dynamic systems and their sensors.

Moreover, in the present chapter two novel approaches for the robust FDI of actuators are developed. The proposed methods are based on the application of the UIF and RUIF to the estimation of the GMDH neural model inputs. Furthermore, the novel approaches of the calculation of the input adaptive thresholds for each input signal of the diagnosed system are developed which enable to perform the robust FDI of the actuators in the dynamic non-linear systems simultaneously. In the case of the RUIF, the fault detection is achieved through the general Unknown Input Observer approach while robustness is attained with the $\mathcal{H}_\infty$ approach. In the usual $\mathcal{H}_\infty$ framework, the prescribed disturbance attenuation level is achieved with the respect to the state estimation error. The proposed approach is designed in such a way that the prescribed disturbance attenuation level is obtained with respect to the input estimation error while guaranteeing the convergence of the observer with a possibly large decay rate of the state estimation error. The proposed design procedure is relatively simple and boils down to solving a set of linear matrix inequalities.

The results of the application of the proposed approach to the identification and robust fault detection and actuators fault localisation of the tunnel furnace are presented. They show that the proposed dynamic non-linear GMDH neural model can be effectively applied to the identification of the tunnel furnace. Moreover, the comparison of the results obtained with the application of the proposed approach and the linear state-space model shows that the GMDH neural model can be successfully applied in the identification of the dynamic non-linear systems tasks. Finally, the resulting neural model is successfully used to design the robust fault detection and localisation scheme.

7

Conclusions and Future Research Directions

Growing demands of reliability and safety of contemporary industrial systems and technological processes impose development of more efficient fault diagnosis methods. The reliable fault diagnosis is expected even in uncertain conditions when measurement noises, disturbances and model uncertainty could appear. The application of the analytical redundancy on the basis of mathematical models opposite to hardware one in the fault detection and isolation systems gave expectations for radical improvement of engineering systems quality. Paradoxically, the application of models in the fault diagnosis systems causes that their effectiveness depends on model quality. In the case of fault diagnosis methods for linear systems it can be assumed that the existing approaches for system identification and fault diagnosis are mature and satisfactory but for non-linear one they are imperfect and require the development. In order to solve such a challenging problem the neural models can be applied. On the ground on the interesting properties the neural models seem to be excellent tools for non-linear systems identification and application for fault detection and isolation. Unfortunately, some weaknesses limit their application in the modern fault diagnosis systems what cause that they should be adopted to the fault diagnosis requirements. Among, the most crucial challenges given for neural models applied in the FDI systems the following can be distinguished:

- Improvement of dynamic neural models quality in order to increase the sensitivity of fault detection systems.
- Development of description of neural models uncertainty allowing to design robust fault detection systems.
- Development of stable dynamic neural models in a state-space representation.
- Integration of neural models and UIFs to design robust FDI systems.

Unfortunately, the classic neural networks often cannot provide acceptable solutions to such difficult problems. Thus, the original objective of this book was to develop efficient neural model-based schemes for solving these

M. Mrugalski, *Advanced Neural Network-Based Computational Schemes for Robust Fault Diagnosis*, Studies in Computational Intelligence 510,
DOI: 10.1007/978-3-319-01547-7_7,

challenging problems in a unified framework. The following is a concise summary of the original contributions provided by this monograph to the state-of-the-art in neural network modeling and fault diagnosis of non-linear dynamic systems:

Improvement of designing methods of dynamic neural models with the application of the GMDH approach. In particular, two models of dynamic neurons are developed i.e., the dynamic neuron with the IIR filter and dynamic neuron in polar coordinates. The synthesis process of the GMDH neural network for the MIMO case is extended. The mechanisms of the network synthesis, which decreases neural model uncertainty, are improved by the introduction of robust parameters estimation methods, development of a new selection method and new neuron quality criterion.

Development of the GMDH model-based robust fault detection methods. In order to achieve this goal novel methods of parametrical uncertainty of dynamical neurons with the application of the LMS, BEA, OBE and Zonotope-based algorithms are developed. It constitutes to the development of methods of calculation of the neural model output uncertainty and subsequently designing output adaptive thresholds allowing to perform robust fault detection.

Development of the MLP model-based robust fault detection method. In order to achieve this goal a novel method of the MLP uncertainty estimation is developed. In such an approach the MLP output uncertainty is obtained on the basis of parametrical uncertainty calculated with the application of the OBE algorithm. Moreover, the algorithm allowing for calculation of the boundary values of the output errors is proposed. Such results allow to obtain the output adaptive thresholds and enable to perform robust fault detection.

Development of the state-space GMDH model-based robust fault detection method. In order to achieve this goal a novel MIMO state-space GMDH neural model is developed. For such a model the conditions of stability of each neuron are defined. The application of the UKF to the constrained parameters estimation allows to obtain an asymptotically stable state-space GMDH neural model. Moreover, the UKF enables to obtain the state-space neural model output uncertainty and calculate the output adaptive thresholds which can be applied to robust fault detection.

Development of the actuator robust fault detection and isolation method. The state-space description of the GMDH neural model allows to apply the UIF and RUIF to the estimation of the model inputs. Such a result enables to develop the input adaptive thresholds for actuators robust fault detection and isolation. In particular, two different methods on the UIF and RUIF are developed. It is especially worth to mention the RUIF method, where the proposed design procedure of an observer-based FDI scheme for which a prescribed disturbance attenuation level is achieved with the respect to the input estimation error while guaranteeing the convergence of the observer with a possibly large decay rate of the state estimation error.

The book also presents a number of illustrative examples and practical implementations of the proposed approaches, which can be summarised as follows:

- Robust FDI of the brushed DC motor via parameters estimation.
- Designing of the MLP model of the valve actuator.
- MLP model-based robust fault detection of the valve actuator.
- Designing of the GMDH neural model of the valve actuator.
- GMDH neural model-based robust fault detection of the valve actuator.
- Designing of the state-space GMDH neural model of the tunnel furnace.
- Robust FDI of the actuators faults of the tunnel furnace on basis of the state-space GMDH neural model and RUIF.

Irrespective of the above results, there still remain open problems regarding some important design issues. What follows is a discussion of the areas proposed for further investigations.

- Development of robust fault isolation methods of sensors faults.
- Integration of the developed neural model-based robust fault diagnosis methods with the fault tolerant control systems.

References

1. Chen, J., Patton, R.J.: Robust Model-based Fault Diagnosis for Dynamic Systems. Kluwer Academic sPublishers, London (1999)
2. Kościelny, J.: Diagnostics of Automated Industrial Processes. Exit, Warsaw (2001) (in Polish)
3. Korbicz, J., Kościelny, J., Kowalczuk, Z., Cholewa, W. (eds.): Fault diagnosis. Models, Artificial Intelligence, Applications. Springer, Berlin (2004)
4. Palade, V., Bocaniala, C., Jain, L.: Computational Intelligence in Fault Diagnosis. Springer, London (2006)
5. Ding, S.: Model-based Fault Diagnosis Techniques: Design Schemes, Algorithms, and Tools. Springer, Heidelberg (2008)
6. Korbicz, J., Kościelny, J. (eds.): Modeling, Diagnostics and Process Control: Implementation in the DiaSter System. Springer, Berlin (2011)
7. Isermann, R.: Fault Diagnosis Applications: Model Based Condition Monitoring, Actuators, Drives, Machinery, Plants, Sensors, and Fault-tolerant Systems. Springer, Berlin (2011)
8. Kościelny, J., Bartyś, M., Syfert, M.: Method of multiple fault isolation in large scale systems. IEEE Transactions on Control Systems Technology 20(5), 1302–1310 (2012)
9. Frank, P.M., Ding, S.X.: Survey of robust residual generation and evaluation methods in observer-based fault detection systems. Journal of Process Control 7(6), 403–424 (1997)
10. Blanke, M., Kinnaert, M., Lunze, J., Staroświecki, M.: Diagnosis and Fault-Tolerant Control. Springer, New York (2003)
11. Patton, R.J., Frank, P.M., Clark, R.N.: Issues of Fault Diagnosis for Dynamic Systems. Springer, Berlin (2000)
12. Uciński, D.: Optimal measurement methods for distributed parameter system identification. Systems and Control Series. CRC Press, Boca Raton (2005)
13. Isermann, R.: Fault diagnosis systems. An introduction from fault detection to fault tolerance. Springer, New York (2006)
14. Witczak, M.: Modelling and Estimation Strategies for Fault Diagnosis of Non-Linear Systems. From Analytical to Soft Computing Approaches. Springer, Berlin (2007)

15. Milanese, M., Norton, J., Piet-Lahanier, H., Walter, E.: Bounding Approaches to System Identification. Plenum Press, New York (1996)
16. Walter, E., Pronzato, L.: Identification of Parametric Models from Experimental Data. Springer, Berlin (1997)
17. Gertler, J.: Fault Detection and Diagnosis in Engineering Systems. Marcel Dekker, New York (1998)
18. Krishnaswami, V., Luh, G., Rizzoni, G.: Nonlinear parity equation based residual generation for diagnosis of automotive engine faults. Control Engineering Practice 3(10), 1385–1392 (1995)
19. Korbicz, J., Bidyuk, P.: State and Parameter Estimation. Digital and Optimal Filtering. Applications. Technical University Press, Zielona Góra (1993)
20. Frank, P., Schreier, G., Garcia, E.: Nonlinear observers for fault detection and isolation. In: Nijmeijer, H., Fossen, T. (eds.) New Directions in Nonlinear Observer Design. Springer, Berlin (1999)
21. Atkinson, A.C., Donev, A.N.: Optimum Experimental Designs. Oxford University Press, New York (1992)
22. Frank, P.M., Marcu, T.: Diagnosis strategies and systems. principles, fuzzy and neural approaches. In: Teodorescu, H., Mlynek, D., Kandel, A., Zimmermann, H. (eds.) Intelligent Systems and Interfaces. Kluwer Academic Publishers, Boston (2000)
23. Soderstrom, T., Stoica, P.: System Identification. Prentice-Hall International, Hemel Hempstead (1989)
24. Ljung, L.: System Identification: Theory for the User. Prentice Hall PTR, Upper Saddle River (1999)
25. Nelles, O.: Non-linear Systems Identification. From Classical Approaches to Neural Networks and Fuzzy Models. Springer, Berlin (2001)
26. Gupta, M., Liang, J., Homma, N.: Static and Dynamic Neural Networks: From Fundamentals to Advanced Theory. Wiley-IEEE Press, Hoboken, New Jersey (2004)
27. Patan, K., Witczak, M., Korbicz, J.: Towards robustness in neural network based fault diagnosis. International Journal of Applied Mathematics and Computer Science 18(4), 443–454 (2008)
28. Rutkowski, L.: Computational intelligence: methods and techniques, Berlin, Heidelberg (2008)
29. Haykin, S.: Neural Networks and Learning Machines. Prentice Hall, New York (2009)
30. Tadeusiewicz, R.: New trends in neurocybernetics. Computer methods in material. Science (2010)
31. Schalkoff, R.: Artificial neural networks. McGraw-Hill Companies (2011)
32. Kacprzyk, J. (ed.): Advances in Intelligent and Soft Computing. Springer, Heidelberg
33. Kacprzyk, J. (ed.): Studies in Computational Intelligence. Springer, Heidelberg
34. Kluska, J.: Analytical methods in fuzzy modeling and control, vol. 241. Springer (2009)
35. Watton, J., Pham, D.T.: An artificial NN based approach to fault diagnosis and classification of fluid power systems. Journal of System and Control Engineering 211(4), 307–317 (1997)
36. Weerasinghe, M., Gomm, J.B., Williams, D.: Neural networks for fault diagnosis of a nuclear fuel processing plant at different operating points. Control Engineering Practice 6(2), 281–289 (1998)

37. Karpenko, M., Sepehri, N., Scuse, D.: Diagnosis of process valve actuator faults using a multilayer neural network. Control Engineering Practice 11(11), 1289–1299 (2003)
38. Fuente, M.J., Saludes, S.: Fault detection and isolation in a non-linear plant via neural networks. In: Proc. 4th IFAC Symp. Fault Detection, Supervision and Safety for Technical Processes, SAFEPROCESS 2000, Budapest, Hungary, vol. 1, pp. 472–477 (2000)
39. Tadeusiewicz, R., Połcik, H.: Applicability of neural models for monitoring and control of selected foundry processes. In: The 5th International Conference Simulation Designing and Control of Foundry Processes, Kraków-Aachen-Sofia (2006)
40. Patan, K., Parisini, T.: Identification of neural dynamic models for fault detection and isolation: the case of a real sugar evaporation process. Journal of Process Control 15(1), 67–79 (2005)
41. Janczak, A.: Identification of nonlinear systems using neural networks and polynomial models: a block-oriented approach, vol. 310. Springer (2004)
42. Palmé, T., Fast, M., Thern, M.: Gas turbine sensor validation through classification with artificial neural networks. Applied Energy 88(11), 3898–3904 (2011)
43. Mozer, M., Smolensky, P.: Skeletonization - a technique for trimming the fat from a network via relevance assessment. In: Touretzky, D. (ed.) Advances in Neural Information Processing Systems 1, pp. 107–115. Morgan Kaufmann, San Mateo (1989)
44. Fahlman, S.E., Lebierre, C.: The cascade-correlation learning architecture. In: Touretzky, D. (ed.) Advances in Neural Information Processing Systems 1, pp. 524–532. Morgan Kaufmann, San Mateo (1990)
45. Mach, M., Golea, M., Rujan, P.: A convergence theorem for sequential learning in two-layer perceptrons. Europhysics Letters (11), 487–492 (1990)
46. Hassibi, B., Stork, D.: Second order derivaties for network prunning: Optimal brain surgeon. In: Touretzky, D. (ed.) Advances in Neural Information Processing Systems 1, pp. 164–171. Morgan Kaufmann, San Mateo (1989)
47. Mezard, M., Nadal, J.P.: Learning feedforward layered networks: the tiling algorithm. Journal of Physics 22, 2191–2204 (1989)
48. Obuchowicz, A.: Optimization of neural network architectures. In: Wilamowski, B., Irwin, J. (eds.) Intelligent Systems: The Industrial Electronics Handbook, 2nd edn., vol. 5, pp. 9–1–9–22. CRC Press, Taylor Francis Group, Boca Raton (2011)
49. Urolagin, S., Prema, K., JayaKrishna, R., Reddy, N.: Multilayer feed-forward artificial neural network integrated with sensitivity based connection pruning method. In: Das, V.V., Stephen, J. (eds.) CNC 2012. LNICST, vol. 108, pp. 68–74. Springer, Heidelberg (2012)
50. Schürmann, J.: Pattern classification: a unified view of statistical and neural approaches. John Wiley & Sons, Inc. (1996)
51. Raudys, S.: Statistical and Neural Classifiers: An integrated approach to design. Springer, New York (2001)
52. Rutkowski, L.: New soft computing techniques for system modeling, pattern classification and image processing, vol. 143. Springer, Heidelberg (2004)
53. Ogiela, M.R., Tadeusiewicz, R.: Modern computational intelligence methods for the interpretation of medical images, vol. 84. Springer, Heidelberg (2008)

54. Su, H., Chong, K.T., Kumar, R.R.: Vibration signal analysis for electrical fault detection of induction machine using neural networks. Neural Computing and Applications 20(2), 183–194 (2011)
55. Tan, W., Nor, N., Bakar, M.A., Ahmad, Z., Sata, S.: Optimum parameters for fault detection and diagnosis system of batch reaction using multiple neural networks. Journal of Loss Prevention in the Process Industries 25(1), 138–141 (2012)
56. Matic, D., Kulic, F., Climente-Alarcon, V., Puche-Panadero, R.: Artificial neural networks broken rotor bars induction motor fault detection. In: 2010 10th Symposium on Neural Network Applications in Electrical Engineering (NEUREL), pp. 49–53 (2010)
57. Ghate, V.N., Dudul, S.V.: Cascade neural-network-based fault classifier for three-phase induction motor. IEEE Transactions on Industrial Electronics 58(5), 1555–1563 (2011)
58. Nozari, H., Banadaki, H., Shoorehdeli, M., Simani, S.: Model-based fault detection and isolation using neural networks: An industrial gas turbine case study. In: 2011 21st International Conference on Systems Engineering (ICSEng), pp. 26–31 (2011)
59. Sina Tayarani-Bathaie, S., Sadough Vanini, Z., Khorasani, K.: Dynamic neural network-based fault diagnosis of gas turbine engines. Neurocomputing (2013)
60. Gillijns, S., Moor, B.D.: Unbiased minimum-variance input and state estimation for linear discrete-time systems. Automatica 43, 111–116 (2007)
61. Mrugalski, M., Witczak, M.: State-space GMDH neural networks for actuator robust fault diagnosis. Advances in Electrical and Computer Engineering 12(3), 65–72 (2012)
62. Witczak, M., Mrugalski, M., Korbicz, J.: Robust sensor and actuator fault diagnosis with GMDH neural networks. In: Rojas, I., Joya, G., Gabestany, J. (eds.) IWANN 2013, Part I. LNCS, vol. 7902, pp. 96–105. Springer, Heidelberg (2013)
63. Isermann, R.: Fault-Diagnosis Systems: An Introduction from Fault Detection to Fault Tolerance. Springer, Heidelberg (2005)
64. Noura, H., Theilliol, D., Ponsart, J., Chamseddine, A.: Fault-tolerant Control Systems: Design and Practical Applications. Springer, London (2009)
65. Theillol, D., Cédric, J., Zhang, Y.: Actuator fault tolerant control design based on reconfigurable reference input. International Journal of Applied Mathematics and Computer Science 18(4), 553–560 (2008)
66. Korbicz, J., Maquin, D., Theilliol, D. (eds.): International Journal of Applied Mathematics and Computer Science: Special issue: advances in control and fault-tolerant systems, vol. 22. University of Zielona Góra Press, Zielona Góra (2012)
67. De Oca, S., Puig, V., Witczak, M., Dziekan, Ł.: Fault-tolerant control strategy for actuator faults using lpv techniques: Application to a two degree of freedom helicopter. International Journal of Applied Mathematics and Computer Science 22(1), 161–171 (2012)
68. Rutkowski, L., Korytkowski, M., Scherer, R., Tadeusiewicz, R., Zadeh, L.A.: Proceedings of the 11th international conference on Artificial Intelligence and Soft Computing. Springer (2012)
69. Farlow, S.J.: Self-Organizing Methods in Modeling GMDH type Algorithms. Marcel Dekker, New York (1984)

70. Ivakhnenko, A., Mueller, J.: Self-organizing of nets of active neurons. System Analysis Modelling Simulation 20, 93–106 (1995)
71. Korbicz, J., Mrugalski, M.: Confidence estimation of GMDH neural networks and its application in fault detection system. International Journal of System Science 39(8), 783–800 (2008)
72. Mrugalski, M., Witczak, M.: Parameter estimation of dynamic GMDH neural networks with the bounded-error technique. Journal of Applied Computer Science 10(1), 77–90 (2002)
73. Patan, K., Korbicz, J., Mrugalski, M.: Artificial neural networks in diagnostic systems. In: Korbicz, J., Kościelny, J., Kowalczuk, Z., Cholewa, W. (eds.) Fault Diagnosis. Models, Artificial Intelligence, Applications, pp. 311–351. WNT, Warsaw (2002)
74. Mrugalski, M., Arinton, E., Korbicz, J.: Dynamic GMDH type neural networks. In: Rutkowski, L., Kacprzyk, J. (eds.) Neural Networks and Soft Computing. Advances in Soft Computing, pp. 698–703 (2003); 6th International Conference on Neural Networks and Soft Computing, Zakopane, Poland, June 11-15 (2002)
75. Korbicz, J., Witczak, M., Patan, K., Janczak, A., Mrugalski, M.: Analytical methods and artificial neural networks in fault diagnosis and modelling of non-linear systems. In: Korbicz, J. (ed.) Measurements Models Systems and Design, Wydawnictwo Komunikacji i Łączności, Warszaw, pp. 175–204 (2007) ISBN: 978-83-206-1644-6
76. Mrugalski, M., Witczak, M., Korbicz, J.: A pole estimation approach to the synthesis of the dynamic GMDH neural networks. In: 2010 Conference on Control and Fault-Tolerant Systems (SysTol), pp. 323–328. IEEE (2010)
77. Mrugalski, M.: An unscented kalman filter in designing dynamic GMDH neural networks for robust fault detection. International Journal of Applied Mathematics and Computer Science 23(1), 157–169 (2013)
78. Widrow, B., Lehr, M.A.: 30 years of adaptive neural networks: perceptron, madaline, and backpropagation. Proceedings of the IEEE 78(9), 1415–1442 (1990)
79. Williams, R.J., Zipser, D.: A learning algorithm for continually running fully recurrent neural networks. Neural Computation 1(2), 270–280 (1989)
80. Tsoi, A., Back, A.D.: Locally recurrent globally feedforward networks: a critical review of architectures. IEEE Transactions on Neural Networks 5(2), 229–239 (1994)
81. Wang, Z., Liu, Y., Liu, X.: State estimation for jumping recurrent neural networks with discrete and distributed delays. Neural Networks 22(1), 41–48 (2009)
82. Kulkarni, R., Venayagamoorthy, G.: Generalized neuron: Feedforward and recurrent architectures. Neural Networks 22(7), 1011–1017 (2009)
83. Duch, W., Korbicz, J., Rutkowski, L., Tadeusiewicz, R.: Biocybernetics and Biomedical Engineeering. Neural Networks 6 (2000)
84. Du, B., Lam, J.: Stability analysis of static recurrent neural networks using delay-partitioning and projection. Neural Networks 22(4), 343–347 (2009)
85. Hunt, K.J., Sbarbaro, D., Zbikowski, R., Gawthrop, P.J.: Neural networks for control systems - a survey. Automatica 28(6), 1083–1112 (1992)
86. Narendra, K.S., Parthasarathy, K.: Identification and control of dynamical systems using neural networks. IEEE Transactions on Neural Networks 1(1), 12–18 (1990)

87. Haykin, S.: Neural Networks. A Comprehensive Foundation. Prientice-Hall, Upper Saddle River (1999)
88. Norgaard, M., Ravn, O., Poulsen, N., Hansen, L.: Neural Networks for Modelling and Control of Dynamic Systems. Springer, London (2000)
89. Elman, J.L.: Finding structure in time. Cognitive Science 14, 179–211 (1990)
90. Parlos, A.G., Chong, K.T., Atiya, A.F.: Application of the recurrent multilayer perceptron in modelling complex process dynamics. IEEE Transactions on Neural Networks 5(2), 255–266 (1994)
91. Karystinos, G.N., Pados, D.A.: On overfitting, generalization, and randomly expanded training sets. IEEE Trans. of Neural Networks 11(5), 1050–1057 (2000)
92. Vapnik, V.N.: An overview of statistical learning theory. IEEE Trans. of Neural Networks 10(5), 988–999 (1999)
93. Hebb, D.: Organization of Behaviour. J. Wiley, New York (1949)
94. Kohonen, T.: Self-organization Maps. Springer, Berlin (1995)
95. LeCun, Y., Denker, J., Solla, S.: Optimal brain damage. In: Touretzky, D. (ed.) Advances in Neural Information Processing Systems 1, pp. 598–605. Morgan Kaufmann, San Mateo (1990)
96. Sarkar, D.: Randomness in generalization ability: A source to impove it. IEEE Trans. of Neural Networks 7(3), 676–685 (1996)
97. Hinton, G.E.: Learning distributed representations of concepts. In: Proc. 8th Annual Conf. of the Cognitive Science Society, Hillsdale, pp. 1–12 (1986)
98. Scalettar, R., Yee, A.: Emergence of grandmother memory in feedforward networks: Learning with noise and forgetfulness. In: Waltz, D., Feldman, J. (eds.) Connectionist Models and Their Implicetions: Readings from Cognitive Science, pp. 309–332. Ablex, Norwood (1988)
99. Kramer, A.H., Sangiowani-Vincentelli, A.: Efficient parallel learning alghoritms for neural networks. In: Touretzky, D. (ed.) Advances in Neural Information Processing Systems 1, pp. 40–48. Morgan Kaufmann, San Mateo (1989)
100. Chauvin, Y.: A back propagation alghoritm with optimal use of hiden units. In: Touretzky, D. (ed.) Advances in Neural Information Processing Systems 1, pp. 519–526. Morgan Kaufmann, San Mateo (1989)
101. Weigend, A.S., Rumelhart, D.E.: The effective dimension of the space of hidden units. In: Int. Joint Conf. Neural Networks: IJCNN 1991, Sattle, USA, pp. 837–841 (1991)
102. Alippi, C., Petracca, R., Piuri, V.: Off-line performance maximization in feedforward neural networks by applying virtual neurons and covariance transformations. In: 1995 IEEE International Symposium onCircuits and Systems, ISCAS 1995, vol. 3, pp. 2197–2200. IEEE, Seattle (1995)
103. Frean, M.: The upstart alghoritm: A method for constructing and training feedforwars neural networks. Neural Computation 2, 198–209 (1990)
104. Ash, T.: Dynamic node creation. Connection Science 1(4), 365–375 (1989)
105. Wang, Z., Di Massimo, C., Tham, M.T., Morris, A.J.: A procedure for determining the topology of multilayer feedforward neural networks. Neural Networks 7(2), 291–300 (1994)
106. Koza, J.R.: Genetic Programming: On the Programming of Computers by Means of Natural Selection. The MIT Press, Cambridge (1992)
107. Marshall, S.J., Harrison, R.F.: Optimization and training of feedforward neural networks by genetic algorithms. In: Int. Conf, Artificial Neural Networks, Boumemouth, UK, pp. 39–42 (1991)

108. Nagao, T., Agui, T., Nagahashi, H.: Structural evolution of neural networks having arbitrary connections by a genetic method. IEICE Trans. Inf. and Syst. E76–D(6), 689–697 (1993)
109. Kosinski, W., Mikolajewski, D.: Genetic algorithms for network optimization. In: Abraham, A., Snásel, V., Wegrzyn-Wolska, K. (eds.) CASoN, pp. 171–176. IEEE Computer Society (2009)
110. Obuchowicz, A., Politowicz, K.: Evolutionary algorithms in optimisation of a multilayer feedforward neural network architecture. In: Proceedings of the Fourth International Symposium Methods and Models in Automation and Robotics, MMAR 1997, vol. 2, pp. 739–743. Institute of Control Engeneering Technical University of Szczecin, Uczelniane Politechniki Szczecińskiej, Szczecin, Wydaw, Międzyzdroje, Polska (1997)
111. Yang, S., Chen, Y.: An evolutionary constructive and pruning algorithm for artificial neural networks and its prediction applications. Neurocomputing 86(0), 140–149 (2012)
112. Doering, A., Galicki, M., Witte, H.: Structure optimization of neural networks with the a*-algorithm. IEEE Transactions on Neural Networks 8(6), 1434–1445 (1997)
113. Obuchowicz, A.: Architecture optimization of a network of dynamic neurons using the a^* algorithm. In: 7th European Congress Intelligent Techniques and Soft Computing, EUFIT 1999, Aachen, Niemcy (1999) CD-ROM
114. Harp, S.A., Samad, T., Guha, A.: Designing application-specific neural networks using genetic algorithms. In: Touretzky, D. (ed.) Advances in Neural Information Processing Systems 1, pp. 447–454. Morgan Kaufmann, San Mateo (1989)
115. Kitano, H.: Designing neural networks using genetic algorithms with graph generation system. Complex System 4, 461–486 (1990)
116. Mrugalski, M., Korbicz, J.: GMDH neural networks. In: Wilamowski, B., Irwin, J. (eds.) Intelligent systems: The Industrial Electronics Handbook, 2nd edn., vol. 5, pp. 8–1–8–21. CRC Press, Taylor Francis Group, Boca Raton (2011)
117. Kondo, T., Ueno, J., Takao, S.: Feedback GMDH-type neural network algorithm using prediction error criterion defined as aic. Intelligent Decision Technologies, 313–322 (2012)
118. Mueller, J., Lemke, F.: Self-organising Data Mining. Libri, Hamburg (2000)
119. Ivakhnenko, A.: Polynominal theory of complex systems. IEEE Transactions on Systems, Man and Cybernetics (4), 364–378 (1971)
120. Mrugalski, M., Arinton, E., Korbicz, J.: Systems identification with GMDH neural networks: a multi-dimensional case. In: Pearson, D., Steele, N., Albrecht, R. (eds.) Artificial Neural Nets and Genetic Algorithms, pp. 115–120. Springer, Vienna (2003)
121. Fasconi, P., Gori, M., Soda, G.: Local feedback multilayered networks. Neural Computation 4(1), 120–130 (1992)
122. Back, A., Tsoi, A.: Fir and iir synapses. a new neural network architectures for time series modelling. Neural Computation 3(3), 375–385 (1991)
123. Gori, M., Bengio, Y., Mori, R.D.: BPS: A learning algorithm for capturing the dynamic nature of speech. In: Proc. Int. Joint Conf. Neural Networks, vol. II, pp. 417–423 (1989)

124. Poddar, P., Unnikrishnan, K.: Nonlinear prediction of speech signals using memory neuron networks. In: Proceedings of the 1991 IEEE Workshop Neural Networks for Signal Processing, pp. 395–404 (1991)
125. Pan, Y., Sung, S., Lee, J.: Data-based construction of feedback-corrected nonlinear prediction model using feedback neural networks. Control Engineering Practice 9(8), 859–867 (2001)
126. Zamarreño, J.M., Vega, P.: State space neural network. properties and appliation. Neural Networks 11(6), 1099–1112 (1998)
127. Lee, T., Jiang, Z.: On uniform global asymptotic stability of nonlinear discrete-time systems with applications. IEEE Trans. Automatic Control 51(10), 1644–1660 (2006)
128. Obuchowicz, A.: Evolutionary Algorithms in Global Optimization and Dynamic System Diagnosis. Monographs, vol. 3. Lubuskie Scientific Society, Zielona Góra (2003) ISBN: 83-88317-02-4
129. Jadav, K., Panchal, M.: Optimizing weights of artificial neural networks using genetic algorithms. International Journal of Advanced Research in Computer Science and Electronics Engineering (IJARCSEE) 1(10), 47 (2012)
130. Cerone, V., Piga, D., Regruto, D.: Bounded error identification of hammerstein systems through sparse polynomial optimization. Automatica (2012)
131. Arablouei, R., Doğançay, K.: Modified quasi-obe algorithm with improved numerical properties. Signal Processing (2012)
132. Alamo, T., Bravo, J., Camacho, E.: Guaranteed state estimation by zonotopes. Automatica 41(6), 1035–1043 (2005)
133. Bravo, J., Alamo, T., Camacho, E.: Bounded error identification of systems with time-varying parameters. IEEE Transactions on Automatic Control 51(7), 1144–1150 (2006)
134. Le, V.T.H., Stoica, C., Alamo, T., Camacho, E., Dumur, D.: Guaranteed state estimation by zonotopes for systems with interval uncertainties. In: Proceedings of the Small Workshop on Interval Methods (2012)
135. Mrugalski, M.: Robust fault diagnosis based on the parameters identification via outer bounding ellipsoid algorithm. In: Malinowski, K., Rutkowski, L. (eds.) Recent Advances in Control and Automation. Challenging Problems of Science: Control and Automation, pp. 418–427. Academic Publishing House EXIT, Warsaw (2008)
136. Mrugalski, M., Korbicz, J.: Parameters estimation methods in the robust fault diagnosis. In: Kowalczuk, Z. (ed.) Diagnosis of Processes and Systems. Control and Computer Science - Information Technology: Control Theory: Fault and System Diagnosis. Pomeranian Science and Technology, pp. 159–166. Publishers PWNT, Gdańsk (2009)
137. Eykhoff, P.: Identyfikacja w układach dynamicznych. PWN, Warszawa (1980)
138. Finigan, B., Rowe, I.H.: Strongly consistent parameter estimation by the introduction of strong instrumental variables. IEEE Trans. Automatic Control AC19(6), 825–830 (1974)
139. Mrugalska, B., Kawecka-Endler, A.: Practical application of product design method robust to disturbances. Human Factors and Ergonomics in Manufacturing & Service Industries (2012)
140. Design and Quality Control of Products Robust to Model Uncertainty and Disturbances
141. Papadopoulos, G., Edawrds, P.J., Murray, A.F.: Confidence estimation methods for neural networks: A practical comparison. IEEE Transactions on Neural Networks 12(6), 1279–1287 (2001)

142. Seber, G., Wild, C.: Nonlinear Regression. John Wiley and Sons, New York (1989)
143. Mo, S.H., Norton, J.P.: Fast and robust algorithm to compute exact polytope parameter bounds. Mathematics and Computers in Simulation 32, 481–493 (1990)
144. Broman, V., Shensa, M.J.: A compact algorithm for the intersection and approximation of n-dimensional polytopes. Math. and Comp. Math. and Comp. in Simulation (32), 469–480 (1990)
145. Dabbenea, F., Gayb, P., Polyak, B.T.: Recursive algorithms for inner ellipsoidal approximation of convex polytopes. Automatica 39, 1773–1781 (2003)
146. Walter, E., Piet-Lahanier, H.: Estimation of parameter bounds from bounded-error data: A survey. Math. and Comp. Math. and Comp. in Simulation (32), 449–468 (1990)
147. Clement, T., Gentil, S.: Recursive membership set estimation for output-error models. Mathematics and Computers in Simulation 32(5-6), 505–513 (1990)
148. Fogel, E., Huang, Y.F.: On the value of information in system identification – bounded noise case. Automatica (18), 229–238 (1982)
149. Maksarov, D.G., Norton, J.P.: Computationally efficient algorithms for state estimation with ellipsoidal approximations. International Journal of Adaptive Control and Signal Processing (16), 411–434 (2002)
150. Dasgupta, S., Huang, Y.-F.: Asymptotically convergent modified recursive least-squares with data-dependent updating and forgetting factor for systems with bounded noise. IEEE Trans. Inf. Theory 33(3), 383–392 (1987)
151. Aarenstrup, R.: Dc motor model (June 2012), `http://www.mathworks.com/matlabcentral/fileexchange/11829-dc-motor-model`
152. Witczak, M., Korbicz, J., Mrugalski, M., Patton, R.: A GMDH neural network based approach to robust fault detection and its application to solve the damadics benchmark problem. Control Engineering Practice 14(6), 671–683 (2006)
153. Mrugalski, M., Korbicz, J., Patton, R.J.: Robust fault detection via GMDH neural networks. In: Proceedings of 16th IFAC World Congress (2005)
154. Mrugalski, M., Korbicz, J.: Application of the mlp neural network to the robust fault detection. In: 6th IFAC Symposium on Fault Detection, Supervision and Safety of Technical Processes - SAFEPROCESS 2006, Beijing, China, vol. 6, pp. 1390–1395 (2006)
155. Mrugalski, M., Witczak, M., Korbicz, J.: Confidence estimation of the multi-layer perceptron and its application in fault detection systems. Engineering Applications of Artificial Intelligence 21(6), 895–906 (2008)
156. Bartyś, M., Patton, R., Syfert, M., de las Heras, S., Quevedo, J.: Introduction to the damadics actuator fdi benchmark study. Control Engineering Practice 14(6), 577–596 (2006)
157. Kościelny, J.M., Bartyś, M., Rzepiejewski, P., Sá da Costa, J.: Actuator fault distinguishability study for the damadics benchmark problem. Control engineering practice 14(6), 645–652 (2006)
158. Mrugalski, M., Korbicz, J.: Least mean square vs. Outer bounding ellipsoid algorithm in confidence estimation of the GMDH neural networks. In: Beliczynski, B., Dzielinski, A., Iwanowski, M., Ribeiro, B. (eds.) ICANNGA 2007. LNCS, vol. 4432, pp. 19–26. Springer, Heidelberg (2007)

159. Korbicz, J., Mrugalski, M., Witczak, M.: On the GMDH approach to robust fault detection of non-linear dynamic systems. In: Busłowicz, M., Malinowski, K. (eds.) Advances in Control Theory and Automation: Monograph Dedicated to Professor Tadeusz Kaczorek on the Occasion of his Eightieth Birthday, pp. 129–140. Printing House of Białystok University of Technology, Białystok (2012) ISBN: 978-83-62582-17-4
160. Mrugalski, M., Korbicz, J.: Uncertainty of dynamic GMDH neural models in robust fault detection systems. In: Grzech, A., Świątek, P., Drapała, J. (eds.) Advances in Systems Science, pp. 211–220. Computer Science. Academic Publishing House EXIT, Warsaw (2010) ISBN: 978-83-60434-77-2
161. Puig, V., Mrugalski, M., Ingimundarson, A., Quevedo, J., Witczak, M., Korbicz, J.: A GMDH neural network based approach to passive robust fault detection using a constraints satisfaction backward test. In: 16th IFAC World Congress, Prague, Czech Republic, [B.m.], [6] CD–ROM (2005)
162. Mrugalski, M.: Robust fault detection using zonotope-based GMDH neural network. In: Korbicz, J., Kowal, M. (eds.) Intelligent Systems in Technical and Medical Diagnostics. AISC, vol. 230, pp. 101–112. Springer, Heidelberg (2013)
163. Clement, T., Gentil, S.: Reformulation of parameter identification with unknown-but-bounded errors. Mathematics and Computers in Simulation 30(3), 257–270 (1988)
164. Fletcher, R.: Practical Methods of Optimization. John Wiley and Sons, Chichester (1981)
165. Back, T., Hammel, U., Schwefel, H.P.: Evolutionary computation: comments on the history and current state. IEEE Transactions on Evolutionary Computation 1(1), 3–17 (1997)
166. Haykin, S.: Kalman Filtering and Neural Networks. John Wiley & Sons, New York (2001)
167. Teixeira, B., Torres, L., Aguirre, L., Bernstein, D.: On unscented kalman filtering with state interval constraints. Journal of Process Control 20(1), 45–57 (2010)
168. Witczak, M., Prętki, P.: Desing of an extended unknown input observer with stochastic robustness techniques and evolutionary algorithms. International Journal of Control 80(5), 749–762 (2007)
169. Gillijns, S., De Moor, B.: Unbiased minimum-variance input and state estimation for linear discrete-time systems. Automatica 43, 111–116 (2007)
170. Nobrega, E., Abdalla, M., Grigoriadis, K.: Robust fault estimation of unceratain systems using an lmi-based approach. International Journal of Robust and Nonlinear Control 18(7), 1657–1680 (2008)
171. Zemouche, A., Boutayeb, M., Bara, G.: Observer for a class of Lipschitz systems with extension to $\mathcal{H}_\infty$ performance analysis. Systems and Control Letters 57(1), 18–27 (2008)
172. Stipanovic, D., Siljak, D.: Robust stability and stabilization of discrete-time non-linear: the lmi approach. International Journal of Control 74(5), 873–879 (2001)
173. Abbaszadeh, M., Marquez, H.: Lmi optimization approach to robust $\mathcal{H}_\infty$ observer design and static output feedback stabilization for non-linear uncertain systems. International Journal of Robust and Nonlinear Control 19(3), 313–340 (2008)

174. Rajamani, R.: Observers for Lipschitz non-linear systems. IEEE Transactions on Automatic Control 43(3), 397–401 (1998)
175. Pertew, A.M., Marquez, H.J., Zhao, Q.: $\mathcal{H}_\infty$ synthesis of unknown input observers for non-linear lipschitz systems. International Journal of Control 78(15), 1155–1165 (2005)
176. Mrugalski, M.: Designing of state-space neural model and its application to robust fault detection. In: Rutkowski, L., Korytkowski, M., Scherer, R., Tadeusiewicz, R., Zadeh, L.A., Zurada, J.M. (eds.) ICAISC 2013, Part I. LNCS, vol. 7894, pp. 140–149. Springer, Heidelberg (2013)
177. Julier, S., Uhlmann, J.: Unscented filtering and nonlinear estimation. Proceedings of the IEEE 92(3), 401–422 (2004)
178. Kandepu, R., Foss, B., Imsland, L.: Applying the unscented kalman filter for nonlinear state estimation. Journal of Process Control 18(7-8), 753–768 (2008)
179. Bernstein, D.: Matrix Mathematics. Princeton University Press, Princeton (2005)
180. Peddle, I.: Discrete state space control. Control Systems 414, 2–3 (2007)
181. Zemouche, A., Boutayeb, M.: Observer design for Lipschitz non-linear systems: the discrete time case. IEEE Trans. Circuits and Systems - II:Express Briefs 53(8), 777–781 (2006)

Index

Printed in the United States
By Bookmasters